T0133628

Design of Integrally-Attached Timber Plate Structures

Design of Integrally-Attached Timber Plate Structures outlines a new design methodology for digitally fabricated spatial timber plate structures, presented with examples from recent construction projects. It proposes an innovative and sustainable design methodology, algorithmic geometry processing, structural optimization, and digital fabrication; technology transfer and construction are formulated and widely discussed.

The methodology relies on integral mechanical attachment whereby the connection between timber plates is established solely through geometric manipulation, without additional connectors, such as nails, screws, dowels, adhesives, or welding. The transdisciplinary design framework for spatial timber plate structures brings together digital architecture, computer science, and structural engineering, covering parametric modeling and architectural computational design, geometry exploration, the digital fabrication assembly of engineered timber panels, numerical simulations, mechanical characterization, design optimization, and performance improvement.

The method is demonstrated through different prototypes, physical models, and three build examples, focusing specifically on the design of the timber-plate roof structure of 23 large span arches called the Annen Headquarters in Luxembourg. This is useful for the architecture, engineering, and construction (AEC) sector and shows how new structural optimization processes can be reinvented through geometrical adaptions to control global and local geometries of complex structures. This text is ideal for structural engineering professionals and architects in both industry and academia, and construction companies.

Yves Weinand is an Associate Professor and Laboratory Director for Timber Construction (IBOIS) at École Polytechnique Fédérale de Lausanne (EPFL), Switzerland. He is the founder of Yves Weinand Architects in Lausanne and Bureau d'études Weinand in Liège, Belgium. Yves Weinand's fundamental research and professional activities investigate the technical possibilities of timber and biobased materials in building technology and construction. Through new innovative approaches and transdisciplinary design methodology, his research ambition is to develop a new generation of renewable and ecological construction.

Design of Integrally-Attached Timber Plate Structures

Yves Weinand

Routledge
Taylor & Francis Group

LONDON AND NEW YORK

First published 2022
by Routledge
2 Park Square, Milton Park, Abingdon, Oxon OX14 4RN

and by Routledge
605 Third Avenue, New York, NY 10158

Routledge is an imprint of the Taylor & Francis Group, an informa business

© 2022 Yves Weinand

British Library Cataloguing-in-Publication Data
A catalogue record for this book is available from the British Library

Library of Congress Cataloging-in-Publication Data
Names: Weinand, Yves, 1963- author.
Title: Design of integrally-attached timber plate structures / Yves Weinand.
Description: First edition. | Abingdon, Oxon ; New York, NY : Routledge, [2022] | Includes bibliographical references and index.
Identifiers: LCCN 2021009876 (print) | LCCN 2021009877 (ebook) | ISBN 9780367689391 (hbk) | ISBN 9780367689384 (pbk) | ISBN 9781003139713 (ebk)
Subjects: LCSH: Wooden-frame buildings--Design and construction. | Framing (Building) | Plates (Engineering)
Classification: LCC TH1101 .W45 2022 (print) | LCC TH1101 (ebook) | DDC 624.1/7765--dc23
LC record available at https://lccn.loc.gov/2021009876
LC ebook record available at https://lccn.loc.gov/2021009877

ISBN: 978-0-367-68939-1 (hbk)
ISBN: 978-0-367-68938-4 (pbk)
ISBN: 978-1-003-13971-3 (ebk)

Typeset in Sabon
by KnowledgeWorks Global Ltd.

I would like to dedicate this book to Louis, Jeanne, and Mila.

Contents

Foreword
Pleasant Matters

Tackling any material in discourse and practice is a difficult task as history shows that materials are often relegated to be dealt with by "industry" in preference to values given to formal questions and conceptual phenomena such as philosophy, ethnography, and geography, which are the main concerns of the architect today. Material is sometimes seen as a focus that will distract the architect from bigger concerns and can therefore be invisible in the discourse. On the other hand, the emphasis on the performance of materials can be found in current theoretical and philosophical discourse, now that architectural discourse has moved away from a linguistic model that limited the role of materials to a singular medium of representation. The climate crisis has allowed a re-examination of this approach and the potential for a new transaction between form and material with critical accounts of the later. Furthermore, in architecture at least, developments in practice support new frameworks in relation to digital-based mechanistic processes such as computer numerical controlled milling operations, rapid prototyping, and advanced finite element analysis allow an immediate link between concepts, production, and actualization, significantly reducing the gaps and distances between these stages. Thinking now takes a new meaning by communicating at levels in matter and thought with far fewer limits, thus, transcending by experimentation, exploration, and discovery of new possibilities for our consideration.

Yves Weinand has written a timely book that connects his forensic knowledge of timber to the conventional boundaries of material science, engineering, computer-aided design, culture, and history to bring art and science back together through a breathtaking tour-de-force of research. Every aspect of the work he has shared with me over a number of years supports our shared explorations of engineering to meet the challenges of the future and has been presented in the book to provide an antidote to the pessimism that exists in practice about the future of timber. More importantly, however, the books expresses the value of persistent work that could help shape the world around us. The book shows how the problem-solving capacity of newfound tools can be put to use in practice and construction to bring prosperity and community through a rational, evidence-based approach developed in his narratives. The spirit of the book is backed by significant scientific research, which in a responsible way resists the blind partisan or traditional defence of wood or an acceptance that wood lacks the inventiveness, novelty and imagination achieved by other materials. The idea behind the book relies on the notion that the forests of the world have the potential to provide a sustainable renewable natural resource. Its basis is that wood products offered by this work could free us from the exhausting forms of what we see and know well. The history and pervasiveness of our association with wood have deep affinities that are complex and beyond their usefulness as a building material.

Yves Weinand makes what is a highly technical subject accessible; timber plate structures and wood-to-wood joints are part of a very finite field and risk the possibility of being relegated to a "fringe" research audience. The book, instead, is a call for action to everyone who is engaged with the built environment. Chapter 3, in particular, takes up the idea of precedence to provide continuity to all audiences through demonstration projects. The examples reinforce the opportunities offered by timber and avoid a fight with the material but conform naturally to the thesis being presented. This is a record of an author who has a deep understanding of timber and the field it sits in with its limitations and potentialities—within science and the built environment—coming with an intense study of a single level of existence of "wood-to-wood" interactions; with temporary and necessary exclusions of problems beyond certain boundaries that are clearly defined. The researchers he has worked in the production of the book add weight to the potential of the book as a reference for future generations.

It is clear to me that we have to raise the issue of how to educate engineers of the future. In that endeavour, the practitioner in me believes we have to begin to redefine the role engineers can play in the defining issues of this Epoch, which are driven by the climate crisis and the newly found power of digital technologies, which should not be the exclusive territory of conventional models of curriculum that standardize what we teach. The work in this book, in my view, begins to show a model of learning that yields more optimistic outcomes. The questions formulated, clearly explained, and answers proposed by Weinand and his colleagues are combined with highly capable self-operated machines with advanced education to build capacities that stand a better chance of coping with the challenges we face in current models of education that increasingly fail to make real and immediate practical impact. Despite the centrality of a particular approach to the use of timber that risks the reduction of architecture to pragmatic operations, on the contrary, this book will stimulate new conversations about the built environment in practice and education opening up a space for discussions about many qualities, uses, and future of timber.

Hanif Kara
February 2021

Preface

The aim of this book is to propose a radically new conceptual design approach toward structures. This new approach is illustrated and applied to innovative timber structures. In detail, the described processes could also be applied to other types of structures and other materials. The conceptual design of structures is at the heart of the design process when the most fundamental and influential decisions are taken for a project. Those decisions occur at an early stage of the design process. At that moment, they need to connect form, space, and structure. Thus, the process and the tools used to be efficient should allow for the unification and the combination of several disciplines, architecture, and structural designs to start with. And indeed, conceptual design approaches to timber structures have always been a fertile ground for multi- or interdisciplinary approaches. The architectural and structural requirements, including details and connections, have to be considered and integrated at a very early stage when it comes to designing timber structures. We intend to generate a new spirit amongst academics and practitioners from engineering, architecture, and other disciplines on how new tools determine conceptual design of structures. The focus is placed on the description of those tools and related case studies who allow for an alternative conceptual design approach embracing form, space, and structure as a holistic whole. This whole also takes into account the new necessity of life cycle assessment and sustainability, which have not been addressed in sufficient manner by structural designers. We believe that the proposed conceptual framework may allow the genius and sensitivity of the designers to find their way more easily into the design process without any historical barriers or preconditions.

Yves Weinand
Liège, January 2021

Acknowledgments

Foremost, I would like to express my sincere gratitude to my former doctoral student and present postdoctoral researcher Aryan Rezaei Rad for his continuous help and input, his motivation, never-ending enthusiasm, and above all, his scientific competence. I could not have imagined having better support for the compilation of this book.

Besides Aryan, I would like to thank my present doctoral student Petras Vestartas for his transdisciplinary and outstanding input and his continuous effort to help and make progress within the IBOIS laboratory.

My sincere thanks go in an equal manner to Andrea Settimi, Alexandre Flamant, and Julien Gamerro, who helped define the book content, figures, and layouts.

My gratitude also goes to my former collaborators at the Laboratory for Timber Constructions (IBOIS), who are extensively quoted throughout this work. Finally, each individual part, which emerged over years of research, appears in a new fashion, is highlighted, and takes on additional meaning within this proposed continuous framework.

Chapter 1

Digitalization in innovative and sustainable timber construction

Yves Weinand

1.1 TECHNOLOGICAL ADVANCEMENTS IN TIMBER CONSTRUCTION

The application of digital-driven tools in the design of spatial timber plate structures embodies both a vision of the future and an understanding of the past. It is inspired by the vision of building as an integrated planning process, where aspects of craft, technique, aesthetic, and structural engineering converge as they did just before the revolutionary 'Age of Enlightenment,' but this time using contemporary engineering methods and tools. This, in particular, demonstrates an act of creativity within the field of architecture and specific contemporary architectural approaches. The raw resource in question has innate qualities that can also satisfy the aesthetic and conceptual designs architects are interested in. The emerging tools in digital architecture, design software, and the digital drawing tool seen as an instrument to conceive architecture, have opened the way for broader digital technology applications, including technical nature. Technical advances that now lie within this context facilitate digital fabrication integration in ways that were unthinkable only a few years ago. Over the last two centuries, the predomination, first of steel, then reinforced concrete within research and applications in civil engineering and materials science, has opened a huge gap of missing research regarding timber as a structural material to be engineered. Our predecessors' and carpenters' intuitive knowledge during the 18th century has been lost with the rise of the engineers who have not taken advantage of timber as a construction material, having *a priori* accorded it a lower level of importance than for steel and concrete.

Technical considerations are often treated as neutral data, which do not, or should not, greatly affect the initial creative design process of a given architect. The technique, construction methods, civil engineering, and static considerations are seen as almost unwelcome ingredients in a certain number of cases. Those supposedly neutral technical considerations are more often than not tackled at a later stage in the design process, compromising the genuinely interdisciplinary and fundamental quality that such research approaches could aspire to.

Timber construction research demands cross-cultural and interdisciplinary approaches involving architecture, civil engineering, and material science. But the timber construction industry itself has remained a particularly conservative and traditional one. The gradual replacement of timber by steel and concrete over the last 200 years has not helped improve new and contemporary timber construction applications from an architectural and civil engineering perspective. This radically new generation of timber structures can change the face of timber construction as an architectural form, both lifting it out of the classical image of traditional architecture and expanding the use of timber in constructions of contemporary character. The old-fashioned image of the "chalet' and related vernacular architecture will be replaced by a contemporary interpretation of timber use in our constructions. It should establish timber as a modern, high-tech material that plays a central role in a society concerned with sustainability.

I.2 INNOVATIVE TIMBER CONSTRUCTION IN THE ARCHITECTURE, ENGINEERING, AND CONSTRUCTION (AEC) SECTOR

Using renewable and sustainable resources in Architecture, Engineering, and Construction (AEC) has become apparent in recent years. Within this context, interest in employing timber as a construction material has increasingly revived. Innovative timber-derived products, such as Laminated Veneer Lumber boards produced with the readily available type of woods, have emerged, and the use of such products is spreading.

This book's ambitious purpose is to develop the next generation of timber constructions made out of innovative engineered timber products at a building scale. It aims at the unprecedented exploration of advanced architectural geometry, engineering analysis, digital fabrication, prototyping, mechanical explorations, and construction of timber structures. Architectural production in recent years has been heavily focused on the digital outcome. With the growing development of computer-aided design (CAD) tools, architects have tried to integrate those new digital tools into their conception. For instance, one such conceptual trend has become known as 'blob' architecture. But the way the architects use the latest digital tools has, until now, had more to do with the role of virtual representation than the role of fabrication. The structural productions derived from the new output of such digital tools have yet to address many formal issues from the digital design revolution on the civil engineering side. Civil engineering structures remain relatively conservative. New research concepts such as 'structural morphogenesis' have emerged recently from within interdisciplinary research environments. Still, the implications for physical and structural investigations constitutes an active part of the form-finding process have not been addressed yet. They need to be explored if such digital tools are to realize their true potential in the fields of architecture, design, and civil engineering.

The research being undertaken at the Laboratory for Timber Construction (IBOIS) at the Swiss Federal Institute of Technology (École Polytechnique Fédérale de Lausanne, EPFL) in Lausanne, Switzerland aims both to explore and to challenge the traditional relationship between engineering sciences and architectural conception. It is ultimately concerned with construction questions in 'real' space, as perceived and used by society. Therefore, we took an entirely different and unique perspective right from the beginning, devoting our attention to exploring in-depth how materialization and physical aspects of 'real' structures are related to their representations in the digital world. We seek to accomplish construction solutions that could be successfully disseminated throughout a construction market, meaning that the realization of unconventional structures at a reasonable cost must be an obligate goal. Developing specific and specialized digital tools appears to be increasingly necessary to pursue the many new questions arising from our ongoing research. A significant part of our approach is concerned with generating and linking together software that works on various levels, from tasks such as the creation of complex shapes to controlling and sizing finite elements and operating computerized numerical control machines (CNCs).

1.3 SUSTAINABILITY ASPECTS AND TRANSDISCIPLINARY DESIGN METHODOLOGY IN ADVANCED TIMBER CONSTRUCTION

Timber is usually thought of as a 'traditional' material, and this is an advantage when it comes to socially legitimating more advanced research into complex shapes and free-form surfaces. Taking an interest in complex geometry from a (timber) construction point of view, instead of only a morphogenetic point of view, is a fundamentally different approach than blob architecture's 'stylized mode' phenomena, and the two should not be confused. Numerous recent buildings designed in the formalistic model of 'blob' architecture show a total lack of sustainable development awareness. This can be seen in the choice of construction materials, how difficult it is to maintain the structure's energy needs, and the high cost of material handling. In contrast, timber as a raw material for construction purposes certainly has a great future in the face of global sustainable development challenges.

In recent years, the necessity of using renewable resources and sustainable solutions in the building sector has become apparent, and timber has been promoted to the center of interest again. New timber-derived products, such as massif block panels, are emerging, and the use of such products has mostly increased over the past years. Their advantages are well-known, especially the low energy consumption for the production of building components (planks, boards, beams, etc.) and research into welded timber could, because of this, lead to exciting industrial opportunities. Savings in time and energy consumption are also notable in timber structure assembly and dismantling processes. However, we have observed that the challenges of sustainable development also concern the issue of architectural form. A fundamental challenge is: How can one integrate a process of formal and technological innovation within a sustainability perspective? A possible solution may lie in rethinking construction techniques and expanding the formal repertoire linked to wood use while affirming the 'traditional' values of timber construction. Together its technical, aesthetic, and environmental appeal can encourage an increase in the use of this material in contemporary construction and set the context through which this book should be understood.

The collaborative approach of architects, civil engineers, mathematicians, and computer scientists in the IBOIS team has offered a unique blend of skills and insights that enabled us to reveal the potential for novel construction applications of a renewable resource. Our strategy of treating morphogenetic aspects and structural aspects on the same level is likely to produce exceptional structural solutions. While the focus of our studies applies specifically to timber, one should also consider the use of other materials and applications. Furthermore, the use of the many small pieces that interact in timber plate structures will be a factor of major importance in determining the probability of structure failure (global failure); the anatomy of timber as a natural, fiber-structured composite should be able to reach higher structural performances when local weak points can no longer affect global stability. This consideration led, in the past, to the invention of plywood. Since plywood is made out of several layers of timber sheets, the sum of those layers is stronger and more rigid or subject to less local failure than the same amount of material taken out of one naturally grown piece of wood. In plywood, the random placement of fiber-perpendicular layers plays a less critical role since they are covered with stronger layers. Taking advantage of this same principle on another scale, we intend to compensate such randomly appearing weak points – contained in a given timber fabric – by a multitude of adjacent and slightly more resistant members who will sustain each other like a fabric using its woven quality as its strength. This will raise the parameter characterizing a value we call 'global failure ratio.' A specific performance factor might be derivable that considers the natural anatomy of timber, which is a disadvantaged construction material in terms of the coefficient of the material defined in Eurocode 5.[1]

With the discussion on global climate change in mind, it is more than obvious that there is a need to change our behavior in many ways.

---

[1] European Committee for Standardisation (CEN) (2008) CEN-EN 1995-1-1:2005+A1 - Eurocode 5: Design of timber structures - Part 1-1: General - Common rules and rules for buildings. Brussels.

Alternative energy resources need to be made accessible and lower consumption of energy achieved. Here, a structure's production and energy consumption play an important role when it comes to timber, a renewable resource, and interesting building material that should be used more frequently. However, environmentally conscious behavior cannot be achieved by obtrusion only. To convince people to invest in environmentally friendly products and materials, these materials have to be attractive. In the past, this was a problem for sustainable architecture's popularity, which was mainly realized with timber as the building material. It has often had a rustic, primitive, and alternative touch, which, though attractive to some, was repellent to many other potential clients. To access the latter group, the design needs to be treated as a strict criterion. The method of contemporary and appealing architecture with timber is both feasible and necessary to become more widely used in construction.

Traditional design and development of structures are primarily based on the concepts of stiffness and efficiency. They are substantially aimed at minimizing the bending of the structural components and generating structures that can be characterized as rigid and inflexible, avoiding elasticity. Disastrous failures of conventional structures (for instance, under conditions of seismic activity, unusually strong wind forces, or unusually heavy snowfall) are reason enough to suggest we must rethink these stiffness and efficiency paradigms to strike a new path in the conception of structures. Here, spatial timber plate construction, especially the way they are deployed, provides exciting perspectives. They offer high-resolution networks made of many individual components. Furthermore, they provide the advantage that singular element failure does not trigger the entire system failure. Such structures admit large deformation without rupture, a property that is highly unconventional for civil engineering structures. The use of such surface elements can also improve safety considerations concerning accidental fire. For instance, a traditional truss can be replaced by a multitude of surface elements (panels) that act socially, like a fabric (as described earlier), retaining the structure's overall integrity even when substantially damaged.

The goal is to improve and expand the uses of timber and timber-derived composites for applications in construction and design. Timber structures made out of simple rectilinear elements have essentially defined timber construction and carpentry for centuries. With new digital tools, timber construction could be transformed, allowing its introduction into a wide range of new applications. As shown by the various digital tools under development, other potentially physically achievable geometries and constructions may emerge first as virtual representations. Such developments introduce a new range of civil engineering challenges of interest in the field of timber construction.

To date, structural analysis has not been widely applied to timber construction as it has to steel or concrete construction. The proposed use of planar structural elements and curved linear elements made out of timber-derived products will introduce timber in constructions, such as public buildings where architectural and aesthetic considerations are deemed to be of strong cultural importance. In the product sector of fills, insulations, and claddings, high-performance, economically attractive ready-to-build-systems, and design objects can be developed from the basic principles described in this book. Other architecture applications are also possible but have not yet been sufficiently developed for industrial applications. For example, one might envisage the industrial production of new classes of timber-derived products, such as woven timber walls and timber composites. Meanwhile, the recent acquisition of a 3D digital scanner allows for precise identification of specific trees and locks and a detailed prescription of their natural anatomy. This enables us to create a link between a particular selection of trees and a specific architectural design or form, avoiding any intermediate geometrical calibration, thus increasing timber profitability by at least 50 percent.

Switzerland has a long tradition in timber construction. The renowned education of carpenters and joiners, the guaranteed craftsmanship, the transfer of know-how from generation to generation, and identifying with the timber as construction and finishing material still build on this long-term experience. The ecological principles and the awareness for the environment (insulation and energy saving) have more and more impact in Switzerland, where the use of timber as a building material is considered an excellent thing to do. The future goal is to use even more timber and timber-based products in residential and commercial buildings. A good variety of timber is available as a local and renewable resource and transferred into high-quality timber products.

Traditional constructions can still be admired, particularly in alpine and rural areas and modern architecture relying on new materials, modern fabrication, and alternative construction methods present in Switzerland. Among other central European countries, Switzerland plays a leading role in the planning and constructing multistory buildings following state-of-the-art design and fire safety concepts, fulfilling the latest earthquake requirements, and using modern (prefabrication) construction methods. New materials and innovative combinations of materials allow for interesting solutions as prefabricated timber concrete composite floors.

Thanks to the 'open' concept of the Swiss design code for timber structures (SIA 265: Timber Structures[2]) and Eurocode 5 for the European Union and the UK, many new ideas could still be realized in prototype solutions, further developed, and finally brought on the market. Recent timber bridges for foot/bicycle traffic, heavy road traffic, and large warehouses and dome structures demonstrate outstanding and state-of-the-art structural timber engineering.

---

[2] Swiss Standard SN 505 265 (SIA 265): Timber Structures.

The quality of the materials and construction today are high and must be maintained. In most cases, integral solutions are offered in terms of modular, prefabricated elements in the field of (multistory) constructions. The connection details, facade systems, windows and doors, thermal insulation, and even interior finish are incorporated into the prefabricated elements before the construction begins on site. In time, productions with rationalized (not necessarily fully automated) production, with immediate delivery on-site for fast construction, have been consolidated. Computer planning, cutting centers, or even robot manufacturing will play an essential role in the different stages of drafting, detailing, fabrication, and planning, minimizing the difference between a one-off design and a serial product. Such techniques make it possible to reintroduce long-forgotten, expensive-to-produce carpentry techniques, such as dovetail connections, which allow for the exact production of any desired architectonic shape and to achieve any prefabrication degree. Given the significant use of timber in construction and a desirable market for timber and timber-based products, there is a vast potential to attract the AEC sector to present their latest engineering, architecture, construction, and research developments.

This book is primarily centered around the use of wood-wood integral connections in timber plate structures. Particular attention is put on the design of form-active surface structures with engineered timber panels. A wide range of geometries ranging from simple configurations to complex forms is essentially investigated. These forms typically lie in the category of spatial structures. In particular, in essence, these structures are an arrangement in 3D space made of planar timber panels. While timber panels can differ in shape and size, they interface with their respective neighboring elements through angular connections. Despite traditional timber-frame structures where timber plates were considered a secondary structural element, in the new design framework for spatial structures, timber plates play a primary role in the load-bearing mechanism.

One of the significant benefits of designing self-supporting, surface-active structures with timber plates is that it offers a sustainable construction by benefiting from the form and geometry rather than material. Computational design and digital fabrication tools have led to the realization of Integrally-Attached Timber Plate (IATP) structures with various shape topologies. Through custom-developed programs, the design and fabrication of structures with many geometrically different plates are possible. For its design, a digital workflow was applied, from the geometry definition to the direct transfer of data for fabrication, to all robotic fabrication of the components. These structures were built following an integrative strategy to combine design, engineering, and fabrication to allow computational feedback and, in particular, to transfer the complex geometry to Finite Element (FE) software and modify the design according to the results. This is achieved by developing custom-scripts generating the machine code for the digital fabrication of the plates and the FE model.

Chapter 2

Structural design methodology in Integrally-Attached Timber Plate structures

Aryan Rezaei Rad and Petras Vestartas

2.1 STRUCTURAL MECHANICS OF WOOD-WOOD CONNECTIONS UNDER DIFFERENT LOAD SITUATIONS

2.1.1 Semi-rigidity in wood-wood connections

Generally, the behavior of connections demonstrates a semirigid performance. In other words, their behavior, shown in Figure 2.1b, lies in between two well-known extremes: ideally pinned behavior (Figure 2.1a) and ideally rigid behavior (Figure 2.1c). The semirigidity of timber connections mainly exists in their flexural behavior, where portions of the active bending moment are transmitted from one timber element to another. This is reflected in the corresponding moment-rotation curves. Nevertheless, the semirigidity of timber connections exists in other kinematics such as tensile and shear behavior, and it is reflected in the corresponding load-deformation curves. This indicates that the interconnected timber elements demonstrate a relative displacement/deformation.

To account for the flexibility in timber connections, their mechanical features should be characterized. Within this context, three main design steps are generally taken regardless of the construction material. Primarily, the kinematic of the connection and the interconnected elements are determined. Determining each interconnected element's position, the relative displacements and rotations of the connection along its principal directions are defined. In the next step, experimental tests are carried out to verify the hypothesis, determine the connection's constitutive behavior, and document the associated force-deformation or moment-rotation curves. A schematic sample of such curves is shown in Figure 2.2a. The curve, in other words, represents the semirigid behavior of the timber connection.

Furthermore, numerical simulations are often employed in parallel to physical experiments to simulate the semirigid behavior. In this phase of the mechanical characterization, design standards are used to identify and classify mechanical characteristics. Performance measures such as design stiffness and resistance (strength), deformations (or rotations), crack

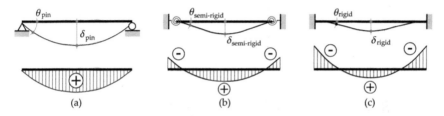

Figure 2.1 (a) Ideally pinned, (b) semirigid, and (c) fully rigid models for connections.

propagation, and initial gap are the main parameters considered in the design process. In the next step, and according to the design objective, the connection's ultimate performance is assessed. This covers the nonlinear behavior of the connection that mainly appears in the postpeak point of the force-deformation behavior. This is a major step to measure the ductility of the connection and classify the corresponding joint region as having brittle, semiductile, or ductile behavior. A spectrum of ductility reflected in the force-deformation is schematically illustrated in Figure 2.2b. In timber structures with carpentry wood-wood connections, fiber orientation of timber board (parallel or perpendicular to load direction), material properties (hardwood, softwood, etc.), and tab insertion angle primarily influence the ductility of connections. Next, and in the last step of the mechanical characterization, the actual load-deformation or moment-rotation behavior of the timber connection is idealized into a multilinear simplified curve. The process is schematically illustrated in Figure 2.2c. This simplified curve represents the constitutive behavior of the connection, and it is used in the analytical or numerical design modeling process. Different methodologies exist in the literature to define such simplified curves for timber connections. In the current study, the European standards EN 26891 [1] and EN 12512 [2] are used to obtain such design curves from the actual behavior of wood-wood connections.

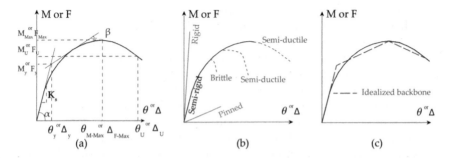

Figure 2.2 Mechanical characterization of timber connections: (a) determination of the force-deformation or moment-rotation behavior (yield, maximum capacity, and ultimate states and slip modulus), (b) performance classification (ductility), and (c) performance idealization.

2.1.2 Kinematics of wood-wood connections

A specific geometry of wood-wood connections, known as one degree of freedom (1DOF) connections, is considered in the current study from a wide range of wood-wood vocabulary. The term 1DOF reflects that only one translational vector in 3D space to (dis)assemble the interconnected elements. The geometry of such connections consists of multiple tabs and slots located on the shared edge of the two intersecting panels. Computing the relative position of the intersecting timber panels, the location of the tabs is first identified, and they are repeatedly inserted in the corresponding slot components. A typical configuration of 1DOF wood-wood connections is shown in Figure 2.3a.

Given that several tab and slots are distributed along the shared edge, this type of connection is also referred to as multiple tab and slot (MTS) joints. The connections' unique geometry dictates the corresponding mechanical kinematics under various loads and allows specific load-transferring mechanisms between the adjacent panels. This will be brought into focus in the following subsections.

Prior to identifying the kinematic of the MTS 1DOF wood-wood connections, it is necessary to understand the assembly logic used for joinery. The assembly of the two panels is directed along the insertion vector, shown in Figure 2.3b. The insertion vector and the vector normal to the locking face are denoted as R and P in Figure 2.3b, respectively. Furthermore, the line segment representing the intersection of the midplane of the two panels is in Figure 2.3a. This line is generally divided into equally spaced segments for the wood-wood connections with multiple tabs, named the tab length L_j.

Given that the vectors n_0 and n_1 are the normal vectors of the two neighboring panels, three principal vectors, denoted as u_1, u_2, and u_3, are consequently defined. These vectors, shown in Figure 2.3b, basically identify the

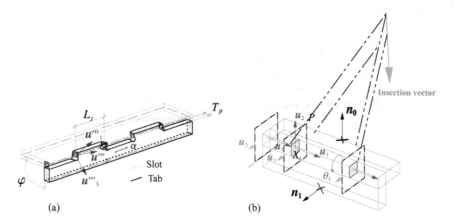

Figure 2.3 Terminology of the multi-tab-and-slot joints. (The geometry of the connections is regenerated based on the existing data and the framework investigated and developed by Roche [3]).

relative translation and rotation between the adjacent panels. It is worth noting that the plane P is normal to the vector $\mathbf{u}_1$, including point, X_i. The geometrical relationship between the three principal vectors and the normal vectors is analytically provided in Eq. 2.1.

$$
\begin{aligned}
\mathbf{u}_1 &= \mathbf{n}_0 \times \mathbf{n}_1 \\
\mathbf{u}_2 &= \mathbf{n}_0 \\
\mathbf{u}_3 &= \mathbf{u}_1 \times \mathbf{u}_2
\end{aligned}
\tag{2.1}
$$

Where $\mathbf{n}_0$ and $\mathbf{n}_1$ are the vectors normal to each timber plate, $\mathbf{u}_1$ is the characteristic vector perpendicular to the plane P, $\mathbf{u}_2$ is the characteristic vector parallel to $\mathbf{n}_0$, and $\mathbf{u}_3$ is the characteristic vector parallel to $\mathbf{n}_1$. The rotation about each of the characteristic angles is defined using the Bryant angles. Accordingly, the rotation about $\mathbf{u}_1$, $\mathbf{u}_2$, and $\mathbf{u}_3$ is defined as angles θ_1, θ_2, and θ_3, respectively.

One of the main challenges in the design of wood-wood connections in timber plates is the fabrication limitation. This constraint mainly refers to the ability of the fabrication tool to mill the timber plate and provide the design geometry associated with the wood-wood connection. The term β_{max} generally expresses the fabrication constraint. This term defines the maximum tool inclination for fabrication, and it is the function of the dihedral angle, φ, as well as the Bryant angles, θ_1, θ_2, and θ_3. The interdependency between the Bryant angles, dihedral angle, and tool inclination controls the possible range and extremes of β_{max}. Given these parameters, three possible combinations generally control β_{max}. In the first scenario, β_{max} is computed according to the diameter of the fabrication tool and the dihedral and Bryant angles. For instance, given a fabrication tool with a diameter of 12 mm, and $\varphi = 120°$ and $\theta_1 = \theta_2 = \theta_3 = 0°$, the maximum tool inclination is determined to be $\beta_{max} = 30°$. This category is schematically shown in Figure 2.4a for an MTS wood-wood connection with dovetail geometry. In the second scenario, β_{max} is computed according to the dihedral angle and θ_3. For instance, given $\varphi = 90°$ and $\theta_3 \leq 30°$, θ_1 and θ_2 are limited to $0°$, and the maximum tool inclination would be $\beta_{max} = 30°$. This category is schematically shown in Figure 2.4b for the same MTS wood-wood connection. In the last scenario, β_{max} is computed according to nonzero Bryant angles. For instance, given that $\theta_1, \theta_2, \theta_3 \neq 0°$ and assuming that $60° \leq \varphi \leq 120°$, the maximum tool inclination would be $\beta_{max} \leq 30°$. This category is schematically shown in Figure 2.4c.

2.2 TIMBER PLATE STRUCTURES WITH WOOD-WOOD CONNECTIONS

This section discusses and studies the application of the wood-wood connection in timber plates and the introduction of a new design framework in spatial complex structures. Employing digital fabrication techniques and information-tool technology, the implementation of the wood-wood connections in timber plate structures aims to improve the sustainability aspects of architecture, engineering, and construction (AEC) while increasing the aesthetic aspects in construction, providing proper structural functionality. Combining the algorithmic computer-aided design (CAD) and traditional carpentry, automatic production of wood-wood connections in thin timber plate structures with nonstandard free-form geometries is introduced. The algorithm is structured such that it provides multiple tabs along the edge of a plate and multiple slots along its mate while it considers both fabrication and assembly constraints. The edgewise connections used to provide the joinery between the neighboring timber plates in a complex assembly enable interlocking of the entire timber plates within the structure without using additional connectors. This mechanism is reached because multiple plates with nonparallel edges are simultaneously connected. Accordingly, the mechanism prevents separation of the structural elements and guarantees a secure force flow among the components. In other words, the geometry of IMAs and timber plates were such that there was no relative movement between the panels after the assembly. Therefore, with the help of the 1DOF assembly concept, a fast and precise sequential assembly is introduced, which immobilizes the geometric arrangement of the timber plates.

Fabricating the wood-wood connections along the edges of timber plates, the 1DOF assembly logic enables a simultaneous assembly of multiple plates. Using this feature, complex and large-scale assemblies are built using segmented components with simple geometry. The Integrally-Attached Timber Plate (IATP) structural system consists of several planar elements with rectangular/parallelogram shapes. For instance, Figure 2.5 shows a doubly curved arch, including two layers of interconnected timber shells. Employing CAD programming interfaces and automatic fabrication technology, the wood-wood connections are fabricated around the perimeter of each timber plate. The system includes the assembly of multiple four-sided hexahedron-shaped boxes, denoted as C_i in Figure 2.5, where $i = 1,2,...,8$. Each box consists of top and bottom plates (denoted as T_i and B_i), a cross longitudinal plate (denoted CL_i), and a cross transverse plate (denoted as CT_i) in Figure 2.5. Within each box, the tenons are located along the edges of the T_i and B_i plates, and their mates (slots) are located in the CL_i and CT_i plates.

Two assembly steps are involved in the construction of the IATP structure shown in Figure 2.5. Four timber plates are assembled to form a timber box (intrabox assembly) in the first step. For this purpose, the CL_i and CT_i plates are first connected via multiple dovetail-shaped joints. This is shown in Figure 2.5, where the CL_1 and CT_1 plates in Box_1 are connected along the vector w_1. Next, and within the first step, the T_i and B_i plates are simultaneously connected to the corresponding CL_i and CT_i plates. For Box_1, this is shown in Figure 2.5, where T_1 and B_1 are connected to CL_1 and CT_1 along the vector v_1. The second step of the assembly corresponds to the assembly of boxes. Within this step, each box is connected to its neighbor

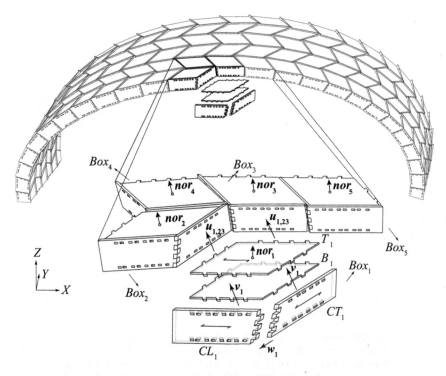

Figure 2.5 Description of Integrally-Attached Timber Plate structure. (The geometry is regenerated based on the existing data and the framework investigated and developed by Nguyen [4] and Robeller et al. [5, 6]).

with only through-tenon joints. Given the IATP structure's geometry in Figure 2.5, each box is simultaneously connected to two other boxes along a single assembly vector. For instance, Box_1 is simultaneously assembled to Box_2 and Box_3 along the vector $u_{1,23}$. Furthermore, each timber box is labeled with a normal vector, identified as nor_i in Figure 2.5. The collective set of normal vectors within a structure can be algorithmically determined to comply with the target surface to achieve the desired geometry.

2.3 PARAMETRIC GEOMETRY GENERATION

2.3.1 NURBS discretization

The nonuniform rational basis spline (NURBS) discretization follows a quad mesh topology considering the assembly sequence of the shells, as shown in Figure 2.6. While the shell structure is based on a two-layer system, the insertion order could be explained using a 2D mesh topology. The NURBS is tessellated by a herringbone quad pattern that reduces each element's mobility to one degree of freedom without colliding parts (see

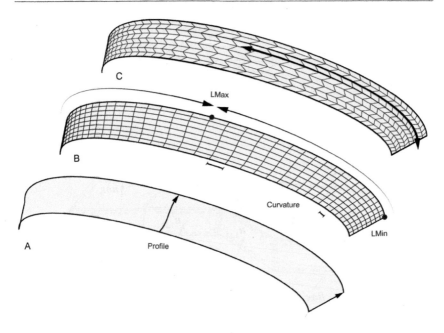

Figure 2.6 NURBS tessellation to quad mesh (a) s-shape profile, (b) Mesh pattern depending on shell curvature, and (c) herringbone pattern to ease assembly sequence and meet fabrication constraints. (The geometry is regenerated based on the existing data and the framework investigated and developed by Robeller et al., 2016 [5]).

Figure 2.6c). There are two advantages of such a pattern: (a) local interlocking (b) global interlocking. Locally, each component blocks previously assembled one. Globally, it results in an interlocking sequence where each row of timber plate components blocks the previous row except for the last group of elements. It allows avoiding long-range escape paths in the assembly that would otherwise be detached simultaneously, i.e., in a regular rectangular grid pattern. Such an assembly sequence is possible due to through-tenon joints that are not perpendicular to the plate edges but rotated to share one common direction per component. Additionally, the rhombus-shaped elements enable to reduce the rotation of the tenon joints to 22.5°, as shown in Figure 2.6c. For the equilateral rectangular, the 45° angle would too larger for the fabrication considering the five-axis cut of mortise degree as shown in Figure 2.6b.

The design surface is a segment following the previously described quad herringbone pattern. The tiling is also used as a graph to construct the two-layer shell structure. The algorithm is applied to the NURBS to generate a basic pattern by evaluating a point grid on the surface. The point-grid density is gradually increased in the main direction, starting with a quadratic 500 mm × 500 mm segment on the ground plane, with a linear increase to

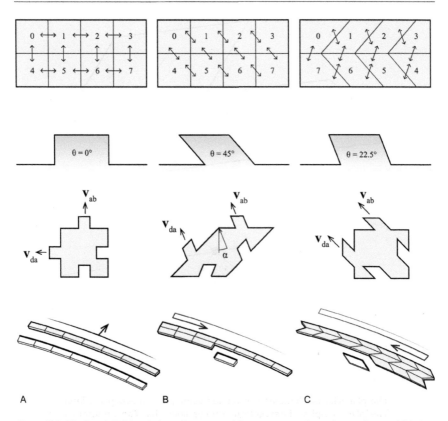

Figure 2.7 Mesh subdivision is based on assembling sequence, insertion order, and fabrication constraints. (a) Quad subdivision with perpendicular joints, (b) angled joints, and (c) shifted pattern. (The geometry is regenerated based on the existing data and the framework investigated and developed by Robeller et al., 2016 [5]).

a maximum segment size of 500 mm × 2500 mm at the top of the shell. The cell differentiates due to curvature changes on the lower end of the shells and the increased loads in this area. These subdivisions result in a quad shell constructed from 312 faces on the first shell and 216 faces on the last one. The mesh edges are indexed following the notation of a component when identifying two sides plates connected to the top and bottom elements. The assembly pattern, insertion vectors, and mesh subdivision are shown in Figure 2.7.

2.3.2 Planarity

The planarization of mesh elements was investigated using two methods: (a) ShapeOP solver (b) plane projection method, as shown in Figure 2.8. The latter is generally used. The underlying quad mesh used for the two-layer

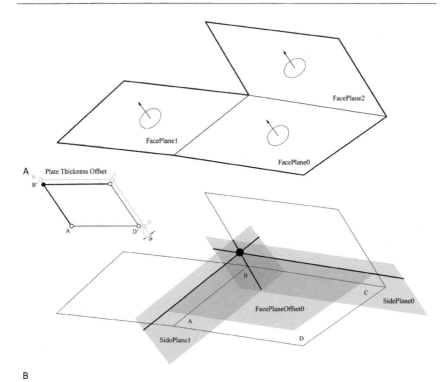

Figure 2.8 The planarity of elements is obtained using (a) plane-to-plane intersections. The plane-to-plate intersection returns lines that can be intersected with a face plane to construct each corner of a plate, and (b) MeshFace plane is obtained by fitting four corners of each mesh quad. (The geometry is regenerated based on the existing data and the framework investigated and developed by Robeller et al., 2016 [5]).

timber plate structure is not represented as a quadrilateral mesh but as a group of disconnected but spatially coupled elements. Elements are not planar within the surface and gradually shift within depending on the NURBS curvature, as shown in Figure 2.9. They face a project to an average plane of a nonplanar quad mesh. This method is applicable to low curvature and high-resolution geometries to keep as large neighboring surface area as possible to satisfy fabrication and assembly objectives.

2.3.3 Existing joinery solver

Joinery Solver was developed because of the past observations and collaborations within researchers of engineering and architecture fields at IBOIS, EPFL. The learning process included global design, local joinery, and fabrication developments by rereading past doctoral theses. The research was focused on state-of-the-art timber structures made from panels. While

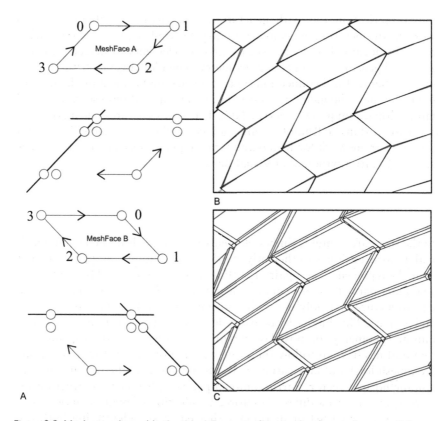

Figure 2.9 Mesh topology (a) the MeshFace winding is the for each row of faces, (b) projected quads to an average plate, (c) volumetric timber plate description following the mesh data structure to perform plane-to-plate intersections. (The geometry is regenerated based on the existing data and the framework investigated and developed by Robeller et al., 2016 [5]).

tools were described as modular and applicable to structures other than pavilion demonstrators, the theoretical descriptions differed from practical software applications. The tools served as a template to design one structural morphology with a limited parametric domain. Consequently, there was a need to create a common geometry library that could help to model timber joinery considering assembly and fabrication constraints.

2.3.4 Contribution to the existing joinery solvers

There are known software applications that allow modeling timber structures considering connections such as Cadwork, Grasshopper plugins (Reindeer, GluLamb, TPS), and other techniques for 3x-axis beam models addressing the purpose of a joinery algorithm as a design tool. Other methods include geometrical considerations but rarely relate to digital-fabrication tool-path generation. The main contribution of this work is an open-source geometry library for timber joints. The model is based on local collisions between pair-wise elements to detect a joinery zone and generate a connection geometry. The global geometry has to be made, whereas local connection types are generated using a predefined tiling technique. The method is tested both using digital representations and fabrication techniques (4.5-axis CNC). The data structures and algorithms are made within the RhinoCommon geometry library and compiled within the .NET framework for Rhino and Grasshopper interfaces and languages such as C#, IronPython, and Visual Basic.

2.3.5 Description of an element for plate representation

In the software context, timber elements have to be described using minimal representation to efficiently model timber joints. A plate element could be represented as a pair of outlines describing top and bottom faces. The plate data-structure contains a pair of outline objects. The outline data-structure is a polyline with additional properties such as an index and a plane. This hierarchy helps to describe a physical timber-plate as a minimal geometry composed of two oriented polygons. Both top and bottom outlines have to follow the same winding order for display methods (i.e., mesh loft) and fabrication methods (i.e., cutting the outline) and its guiding normal vector. Furthermore, a plate can belong to a larger group of components or modules; therefore, a collection of plates can be represented as a sorted list or dictionary that has a key identifying both the hierarchy and the individual element. The plate edges include vectors to specify the insertion direction of joints and one common insertion vector per plate. A plate contains cuts to establish connections needed to cut it out from closed planar outlines using curve intersection or solid Boolean operations.

2.3.6 Groups of minimal models – SortedList

It is rarely possible to assemble a model considering only singular elements. Several terms were used before, such as *boxes*, *components*, and *sections*, which are used to indicate that a model is composed of nested singular elements. The main question is to understand the level of nesting and the way it is implemented. Computation of a SortedList (dictionaries in general) helps to keep a geometrical system linear, where keys give information about grouping. For instance, a tag (A; B) would show that a plate A belongs to a box group B, and the tag could be as long as needed (A; B; C; D, ...n). Rhino and Grasshopper users are familiar with the concept of a DataTree that is a similar implementation for the new geometry development of timber joints.

2.3.7 Detection of joinery within a global timber plate model

Parametric models for tessellated NURBS are commonly used to obtain geometries for timber plate structures. However, such models only serve in case studies. Also, such models are relatively fast due to dependency on a mesh data structure. Such models have to be revised to have a flexible system that can be reused. A graph structure of interconnected plates could be made by performing distance searches of plate axes and faces. Given the two joint types, edgewise dovetail and tenon-mortises, the joinery solver can detect different connections by location of a side and end of a plate. If a plate is connected side-to-top or bottom face, then a joint is considered a side-end, and if plates are facing end-to-end, then a dovetail connection is applied. A similar process can be applied to other connection types, i.e., side-side, half-lap, etc. The iteration of plates and closest neighbor search can be slow because each plate must be checked against the rest of the elements. It is possible to decrease search time by introducing a bounding volume hierarchy or performing spatial indexing such as RTree. Overall, the search methods help identify connection morphology used for constructing connectivity graphs, deriving assembly sequence, and being used as a primary input for joinery geometry generation.

2.3.8 Local connection model

Joints are commonly described in the literature using small scale drawings of pair-wise joints without a global definition of a structural system. The Joinery Solver starts from such a description by introducing a Tile class, a data structure representing a timber joint inscribed in a Unit (1x1x1) box. The tile contains male and female cuts. This local model could be transformed using shear matrices to adapt to various angles. The assumption is made that ChangeBasis matrix transformation and a translation matrix could be applied to adjust to multiple angles of a joint. A Cut data structure represents the female and male cutting outlines in a Tile. The Cut contains information about machining methods such as Mill, Cut, Drill, SawBlade, etc. Also, machining parameters include tool radius, notch orientation, and visualization geometry for polyline or BRep Boolean operations. A tile set is finite by grouping connections into end-end, side-side, side-end, half-lap, end, etc. Each group of tiles may contain subcategories depending on the design of a joint. There are two methods to describe a joint: (a) the developer creates a list of functions for commonly used connections or (b) a user can create a customized joint required for a project study. This joint set is needed to compute timber joinery for a global geometry and adapt to a local connection model.

2.3.9 Generating joints on connection zones

Joints could be generated when connection zones are identified, and local joints are described. First, a connectivity graph is iterated to know what function has to be applied to a pair of elements (i.e., side-side, side-end, end-end, etc.). Second, geometrical operations are made based on different categories to obtain the top and bottom outlines representing a transformed box in a global model. Third, a list of tiles is iterated per connection node because a joint can contain a joint group, such as a half-lap joint with a screw. Fourth, a ChangeBasis and Translation matrices are applied to transform a tile from the Unit-Box to 3D. The tiling process can be based on precomputed tiles that work faster because it is unnecessary to recompute each time a joint has to be applied to a plate, except in cases where joint parameters differ, such as division of tenons based on an edge length. Lastly, the plates and their cuts are returned for visualization and fabrication.

2.3.10 Polyline Boolean intersections

In geometry context, joints could be cut-out from a timber element performing a Boolean Difference operation. The method is slow and not reliable for most CAD applications (RhinoCommon). However, regular sections such as plates and straight beams could apply a 2D curve line-line intersection or polyline insertion into another polyline. The point-inclusion and closest distance to polyline methods are performed to know which part of a joint polygon is inside the plate outline. The largest part is taken from the plate or beam polygon and is merged with the joint segment. The same process is repeated for all sides of the polygons with a joint and top and bottom outlines. The contour could be meshed by applying an ear-clipping algorithm to get a 3D representation of a plate and edgewise joints.

2.3.11 BRep Boolean intersections

The Boolean operations cannot be avoided for some timber structures. In the fabrication context, a joint could be represented as a 3D volume that has to be subtracted from a timber element. For design and development, these volumes could be previewed without performing Boolean operations. For final design and fabrication, the slower process could be applied (up to several seconds per element). Several considerations need to be addressed: floating-point errors fail the Boolean operations (RhinoCommon) and non-manifold conditions when joints are touching face-to-face, edge-to-edge, and vertex-to-vertex would not work. Currently, the application does not offer any solution besides possible implementation of the state-of-art work in research that needs to be incorporated into common CAD applications. These points must be taken into account while developing an algorithm for a timber structure using wood-wood connections.

2.3.12 Orienting and nesting and CNC tool-path

Shell structures made from multiple self-similar elements present a question: how should we order elements to produce efficient material during fabrication and maintain order for assembly? The algorithm used to generate spatial segmented timber plate shells contains outlines representing the top and bottom polylines and planes for orientation from 3D space to the XY plane. Furthermore, elements have to be oriented based on sequential component indexing. Additionally, plates have to be placed in a grid-like pattern on a 4 cm laminated veneer lumber (LVL) board or nested to minimize waste using packing algorithms such as OpenNest. The same procedure has to be applied for all arches and their components while running CNC fabrication simultaneously.

2.4 MECHANICAL STUDIES: TIMBER PLATE STRUCTURES WITH WOOD-WOOD CONNECTIONS

To gain insight into the global performance of IATP structures, the behavior of the connections should be researched first. The wood-wood connections can potentially demonstrate a semirigid behavior along and/or about each characteristic vector (u_1, u_2, and u_3) shown in Figure 2.3. The geometry of a single wood-wood connection with a through-tenon is shown in Figure 2.10. The translational tensile, edgewise (in-plane), and flatwise (out-of-plane) behavior of the joint is labeled as axes 1 to 3 in Figure 2.10. Furthermore, given the Bryant angles shown in Figure 2.3 (θ_1, θ_2, and θ_3), the torsional, out-of-plane flexural, and in-plane flexural behavior of the joint are labeled as axes 4 to 6 in Figure 2.10, respectively. The following sections investigate the kinematic of such wood-wood connections under four main load cases: tensile loads (axis 1 in Figure 2.10), edgewise loads (axis 2 in Figure 2.10), flatwise loads (axis 3 in Figure 2.10), and flexural moments (axis 5 in Figure 2.10).

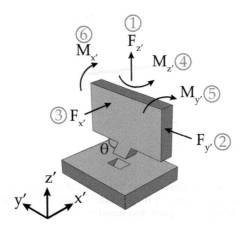

Figure 2.10 Generic degrees of freedom (DOF) defined for through-tenon joints.

2.4.1 Tensile behavior[1]

This section discusses the behavior of the wood-wood connection under tensile loads (DOF 1 in Figure 2.10). Although two different assemblies are considered, the tensile load-carrying mechanism of the wood-wood connection remains the same. In the first assembly, an IATP box component is studied (Figure 2.11). Within the box, the insertion vector denoted as $V_{i,int}$ indicates the assembly direction between the interconnected plates. Given the edges of the plate T_i, denoted as $Edge_{i,1}$ and $Edge_{i,2}$ in Figure 2.11, the vectors $V_{i,edge,1}$ and $V_{i,edge,2}$ indicate the direction of the tensile load applied to the wood-wood connections. Generally, it is assumed that $V_{i,edge,j}$ is normal to $Edge_{i,j}$. Furthermore, the tensile behavior of the wood-wood connection lies within the local x'y' plane. It is worth noting that the tab insertion angle of the wood-wood connection (defined as θ in Figure 2.10) should not be 90°. Otherwise, the connection would not have any axial resistance along $V_{i,in}$ and the tenon part would recede from the slot part without any resistance.

The second assembly used to demonstrate the tensile load-carrying mechanism of the wood-wood connections corresponds to three IATP box components that are simultaneously assembled. This is shown in Figure 2.12, where C_i, C_{i-1}, and C_{i-2} are assembled. Similar to the previous assembly, $V_{i,ext}$ indicates the insertion path that joints C_i to C_{i-1} and C_{i-2} using the wood-wood connections of the T_i and B_i plates. Furthermore, the edges of the T_i plate are denoted as $Edge_{i,1}$ and $Edge_{i,2}$, and they are parallel to the edges of the neighboring plates denoted as $Edge_{i,i-1}$, and $Edge_{i,i-2}$, respectively.

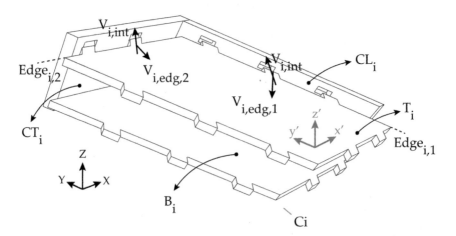

Figure 2.11 Tensile load-carrying mechanism between the plates T_i and CL_i and the plates T_i and CT_i. (The geometry is regenerated based on the existing data and the framework investigated and developed by Rezaei Rad et al., 2019 [8]).

---

[1] Research credit: A. Rezaei Rad, 2020 [49]

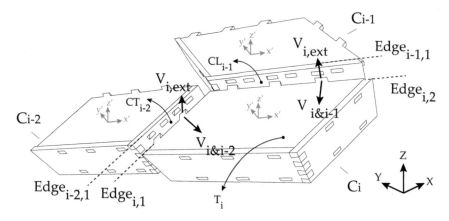

Figure 2.12 Tensile load-carrying mechanism for interbox assembly. (The geometry is regenerated based on the existing data and the framework investigated and developed by Rezaei Rad et al., 2019 [8]).

Accordingly, the direction of the tensile load-carrying mechanism between C_i and C_{i-1} is assumed perpendicular to $Edge_{i,1}$ and denoted as $V_{i\&i-1}$. Moreover, the direction of the tensile load-carrying mechanism between C_i and C_{i-2} is assumed perpendicular to $Edge_{i,2}$ and denoted as $V_{i\&i-2}$. Similar to the previous assembly, the tensile load-resisting mechanism rests in the local $x'y'$ plane of the T_i plate.

2.4.2 Edgewise behavior[2]

This section researches the behavior of the wood-wood connection under edgewise (in-plane) loads (DOF 2 in Figure 2.10). The edgewise loads are transmitted within the local $x'y'$ plane of each timber plate, and therefore, they are also referred to as in-plane loads. In other words, the in-plane loads are applied parallel to the edges of timber plates. The term 'edgewise' is used to refer to this force-flow mechanism. The edgewise load-carrying mechanism associated with an intrabox and an interbox assembly and the local coordinate system corresponding to the T_i plates is shown in Figure 2.13. The wood-wood connections identified in Figure 2.13a are responsible for keeping the in-plane consistency between the boxes C_i and C_{i-1} by resisting the loads that act in the $x'_i y'_i$ plane and along the vector $V_{i,i-1}$. This vector is parallel to the edge of the top plate in C_i. Similarly, the wood-wood connections identified in Figure 2.13b maintain the in-plane consistency between the C_i and C_{i-2} boxes by resisting the loads that act in the $x'_i y'_i$ plane and along the vector $V_{i,i-2}$. This mechanism is repeated in all IATP boxes, where the TT joints identified in Figure 2.13c carry the loads acting in the local $x'_i y'_i$ plane and along the V_i and V_j vectors.

---

[2] Research credit: A. Rezaei Rad, 2020 [49]

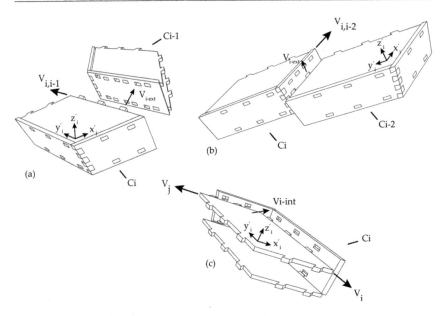

Figure 2.13 **Edgewise load carrying mechanism of the through-tenon wood-wood con-nections: (a) Across the C_8 and C_5 boxes, (b) across the C_8 and C_7 boxes, (c) within a single box. (The geometry is regenerated based on the existing data and the framework investigated and developed by Rezaei Rad et al., 2019 [7]).**

2.4.3 Flatwise behavior[3]

This section researches the behavior of the wood-wood connection under flatwise (out-of-plane) loads (DOF 3 in Figure 2.10). Similar to the previous kinematics, the flatwise behavior is studied for an inter- and intrabox assembly, as shown in Figure 2.14a-b, respectively. For the intrabox assembly (Figure 2.14a), $Vec_{i,int}$ denotes the insertion vec-tor used to assembly the top and cross plates. For the interbox assembly (Figure 2.14b), $Vec_{i,ext}$ denotes the insertion vector used to assembly the C_i box to the C_{i-1} and C_{i-2} boxes. The local coordinate system associated with the top plate of C_i per each assembly is identified as x', y', and z'. For the intrabox and interbox assembly, the flatwise behavior is defined such that the neighboring plates relatively deform along the vector $u_{z'}$ and $u_{Ci,z'}$, respectively. In other words, the direction of the flatwise behavior is parallel to the local z' axis (perpendicular to the local x'y' plane) for the T_i plate. Given that the flatwise loads are transmitted along the direction normal to the local x'y' plane for each timber plate, they are also referred to as out-of-plane loads.

---

[3] Research credit: A. Rezaei Rad, 2020 [49]

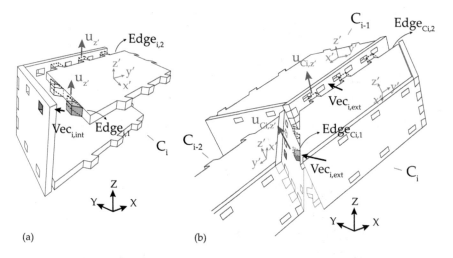

Figure 2.14 Out-of-plane load-carrying mechanism for an (a) interbox assembly, (b) intrabox assembly. (The geometry is regenerated based on the existing data and the framework investigated and developed by Rezaei Rad et al., 2020 [9]).

2.4.4 Flexural behavior[4]

The rotational behavior of the wood-wood connections focuses on characterizing the response of the connection under out-of-plane flexural moments (DOF 5 in Figure 2.10). The geometry of a wood-wood connection and its local coordinate system is shown in Figure 2.15i-j. To describe the moment-rotation behavior, the intra- and interbox assembly of timber plates are shown in Figure 2.15a-b, respectively. The flexural moment aims to rotate the joint about the edges of the T_i plate for the intra- and interbox assembly. In detail, the connections under the flexural moment relatively rotate about the vectors $V_{edge, i}$ and $V_{edge, j}$ as shown in Figure 2.15i. These vectors represent the direction of the plate edges, and they are also shown in Figure 2.15a-b. Furthermore, the flexural moment is such that the tenon and slot components of the wood-wood connection relatively rotate about the y' axis (Figure 2.15i). This indicates the out-of-plane rotation, and therefore, the kinematic is recognized as out-of-plane flexural behavior. Figure 2.15j shows the rotation axis of the joint and associated undeformed shape. It is worth noting that the flexural behavior about the local x' and z' axes is deemed to be rigid. These kinematics are identified by springs #4 and #6 in Figure 2.10. For numerical simulations, an infinitely rigid value is assigned to these springs to reflect their rigid behavior.

---

[4] Research credit: A. Rezaei Rad, 2020 [49]

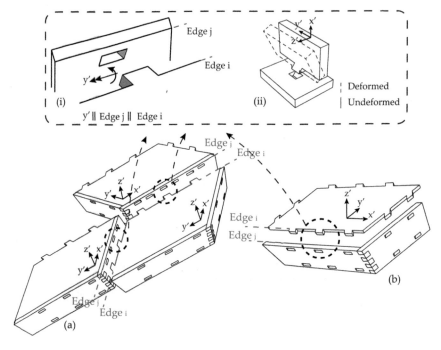

Figure 2.15 **Load-carrying mechanism of the wood-wood joints under flexural moments.** (The geometry is regenerated based on the existing data and the framework investigatedand developed by Rezaei Rad, 2020 [49]).

2.5 EXPERIMENTAL STUDIES

The physical behavior of the through-tenon wood-wood connections is complex. This is mainly attributed to the direct contact between the tenon and slot components and associated interaction, an interlocking mechanism, complex force flow, and the geometrical relationship among timber plates in 3D space. Connections substantially influence the global behavior of timber structures and are, therefore, one of their key components. In particular, the semirigidity of the connections has to be taken into account in the model since simplifications considering the connections as rigid or hinged lead to under- and overestimated displacements, respectively. Given that Through-Tenon joints are the primary connection mechanism in IATP structures, an experimental model is developed to have a clear understanding of their behavior. The design parameters, including the dimension of the joints and the material properties of the timber connection, are first introduced. Next, the test setup associated with the load cases that were previously introduced in Section 3 is presented. The experimental testing procedure is explained, and the quantitative performance assessment of the timber connections is accordingly discussed.

2.5.1 Design parameters: Geometrical and material properties of the wood-wood connections

Different geometrical and mechanical parameters affect the design and performance of the Through-tenon wood-wood connections. In particular, fiber orientation of timber panels, tab insertion angle of the tenon components, material properties, number of plywood layers per timber panel and their arrangement (distribution) over the panel thickness, moisture content of timber, the dimension of the fabrication tool used for milling, and the dimension of tenons and slots are the most critical design parameters. While Rezaei Rad, et al. [7, 8, 9] and Roche et al. [3, 10, 11] widely investigated the influence of each of the design parameters as mentioned above, the focus in the current book is particularly put on a specific through-tenon joint. Accordingly, the design space is considerably reduced, and the detailed behavior of the timber connection of interest is thoroughly investigated. The following two paragraphs explain the geometrical, material, and mechanical features of the Through-Tenon wood-wood connections.

Beech BauBuche LVL panels made out of hardwood material supplied by Pollmeier were used as the construction material [12, 13]. Each panel had 13 layers of 3.3 mm-thick plywood, forming a 40 mm-thick timber plate [14]. The composition of the layers along the section is |||—|||||—|||, where '|' stands for the longitudinal and '—' stands for the cross-layer. Given the local axes defined for each timber plate in Figure 2.16a, the timber panels are oriented such that the main fibers of the interconnected tenon component of the joint are oriented perpendicular to the local y' axes in the current study. Accordingly, 11 crosswise and two longitudinal fiber layers are included over the tenon cross section (cross-plies ratio = 11/13). Figure 2.16b shows the distribution and configuration of the fiber layers.

With respect to the geometrical properties of the connection, tab insertion angle and the connection length are the essential design parameters controlled by fabrication, assembly, and eventually engineering constraints. The tab insertion angle is primarily controlled by the fabrication tool and milling constraints. In fact, the maximum amount of inclination possible for the tab element is dictated by the permissible degree of rotation in the five-axis CNC machine. Using a five-axis CNC machine with a fabrication tool with 20 mm in diameter, the tab insertion angle used to assemble the tenon and slot components was set to be 60°. Regarding the length of the tenon element, this parameter is dictated by the assembly and engineering constraints. One of the constraints in the assembly of IATPs, especially in large-scale structures, is to avoid overlapping between the TT joints. Therefore, the design process should ensure that an appropriate tenon length is adopted.

A limited number of tenons (and slots) can be allocated around the edges of timber plates and a relevant length should be adopted for the connection to ensure structural functionality. Given these two constraints, the tab

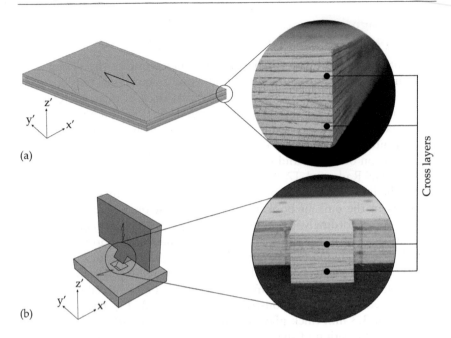

Cross layers

(a)

(b)

Figure 2.16 (a) Beech BauBuche LVL panel and the distribution of veneer layers over the cross section [12, 14, 13], (b) configuration of the through-tenon wood-wood connections concerning the fiber orientation.

length is set to 50 mm. This length is suggested by the results obtained from preliminary numerical simulations. It is also worth noting that the joints can potentially interact with each other. This interaction depends on the tenon length, as well as the number of joints at each edge. In the current study, it is assumed that the joint-spacing is large enough that we can be reasonably assume that the interaction is minimal and can be neglected. Other geometrical properties of the joint are the tab height and width. In the current experimental model, the tab height, including the volume in contact with the slot component and the protrusion, is 60 mm, and the tab width is the same as the thickness of the timber boards (40 mm). Eventually, the joints' geometry is generated with an algorithmic tool, and the associated G-code is introduced to a robot router for digital fabrication. Figure 2.17 shows the geometrical properties and configuration of the wood-wood connection, together with the fiber orientation of the tenon and slot components.

2.5.2 Test setup and associated instrumental of the wood-wood connection

The dimension of the specimens, including the tenon and slot components, steel supports, and the instrumentation of the experimental test setup used to investigate the behavior of the through-tenon wood-wood connection

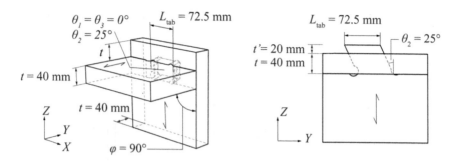

Figure 2.17 Geometrical properties of the through-tenon wood-wood connection used in the experimental study.

under tensile loads, are illustrated in Figure 2.18. By aligning the center of the rigidity of the connection area with the tensile loading direction (Figure 2.18, A-A), the test specimen is designed to have a symmetric geometry. Accordingly, the tensile force-deformation behavior is isolated, and the influence of undesirable effects (i.e., shear loads and bending moments) is minimized. Consequently, it is deemed that the design test setup can simulate the tensile kinematic behavior of the wood-wood through-tenon connections described in 2.4. Fixing the slot component to the hydraulic machine by means of a steel plate, the tensile load is applied to the slot component while the tenon component is fully fixed to rigid support by means of two L-shaped steel supports. The relative deformation between the slot and tenon elements are measured using linear variable differential transducers (LVDTs). Two LVDTs installed on the ground are attached to the right and left sides of the slot component. These LVDTs are shown in Figure 2.18, C-C.

The dimension of the timber specimens and the typical instrumentation of the experimental test setup used to investigate the behavior of the Through-Tenon wood-wood connection under edgewise loads are illustrated in Figure 2.19. The primary focus in the design of the test setup was to minimize the influence of undesirable factors. To do so, a symmetric configuration is designed for the geometry of the test specimen, including three plate elements (Figure 2.19). Furthermore, the surface of the slot components was in full contact with the edge of the tenon. The advantage of the two latter assumptions is twofold. First, it ensures that unwanted loading mechanisms, such as torsion, are less prone to occur. Second, the test setup can uniformly distribute the in-plane loads between the two wood-wood connections and avoid in-plane flexures and rotations. Consequently, a force-deformation response that realistically reflects the in-plane (edgewise) behavior of the wood-wood connection in large-scaled structures is obtained.

After assembling the slot and tenon components, the two slot components were placed on top of a rigid concrete block, and the hydraulic jack applied the in-plane load to the slot component at its top. The designed

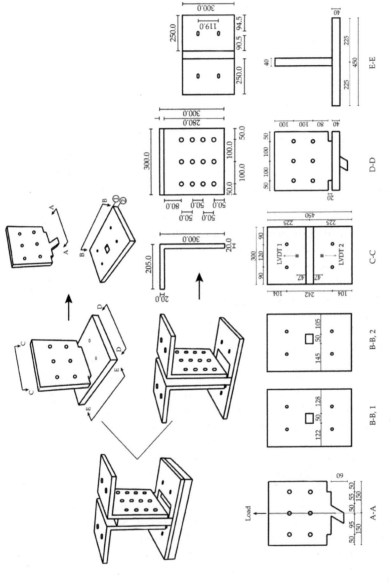

Figure 2.18 The dimension of the specimen and design test setup for the connections under tensile loads.

Image credit: A. Rezaei Rad, 2020 [49]

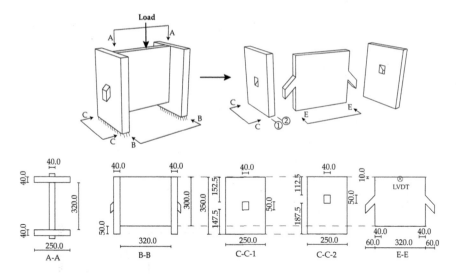

Figure 2.19 The dimension of the specimen and design test setup for the connections under edgewise loads.

Image credit: A. Rezaei Rad, 2020 [49]

setup is similar to the one investigated by Hassanieh et al [15, 16]. One LVDT device was installed on the top-middle side of the tenon component, as shown in Figure 2.19 E-E. Similar to the previous load case, the LVDT was attached to the ground. Moreover, the slot components were fully attached to the concrete block. Thus, the LVDT was used to measure the displacements of the tenon component relative to the base. This included the relative deformation between the tenon and slot component and the deformation caused by compression strains in the slot component.

The dimension of the timber specimens and the typical instrumentation of the experimental test setup used to investigate the behavior of the Through-Tenon wood-wood connection under edgewise loads are illustrated in Figure 2.20. Similar to the edgewise load cases, a symmetric configuration was adopted by aligning the center of the rigidity of the tenon component with the direction of the out-of-plane load in order to minimize undesirable effects and unwanted loading. Since the tab insertion angle is 60°, the rotation of the tenon component at its four corners is not equal. Therefore, each corner of the tenon plate is equipped with one LVDT device to measure the relative deformation between the slot and tenon elements. The exact position of the LVDTs in the tenon component is shown in Figure 2.20.

After assembling the slot and tenon components, the two slot components were placed on top of a rigid concrete block. Four u-shaped steel pieces, shown in Figure 2.20, were used at the top and bottom of the test specimen to guarantee that the slots components remain plumb (vertical) during the loading process. These plates were mainly responsible for preventing the test specimens

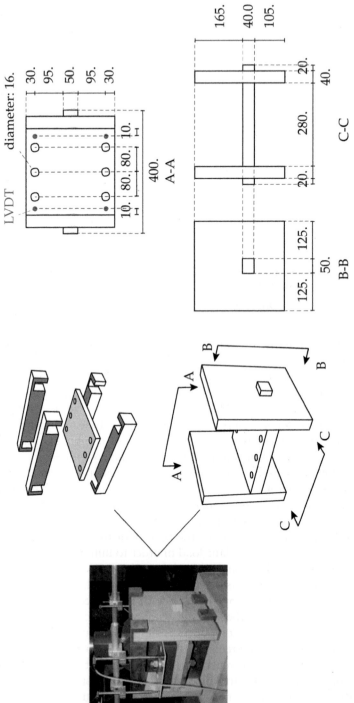

Figure 2.20 Dimension of the specimen and design test setup for the connections under flatwise loads.

Image credit: A. Rezaei Rad, 2020 [49]

from bending and ensuring that the test specimen is exclusively subjected to out-of-plane translational forces. Furthermore, a rigid rectangular steel plate is attached to the bottom surface of the tenon component. The idea was to prevent the tenon component from bending and ensuring that the surface load is uniformly distributed to the tenon component. Next, the flatwise load was applied by a 300 kN hydraulic jack to the tenon component at its top.

The dimension of the timber specimens and the typical instrumentation of the experimental test setup used to investigate the behavior of the Through-Tenon wood-wood connection under flexural moments are illustrated in Figure 2.21a and Figure 2.21f-g. The test setup employed in this investigation was primarily designed by Roche [3]. The platform is designed to isolate the out-of-plane flexural behavior and minimize undesirable effects associated with unwanted loadings such as torsion and shear.

The tenon and slot components are first assembled (Figure 2.21d-e). The specimen is then mounted on rigid support shown in grey in Figure 2.21a. A lever arm between the specimen and the hydraulic jack is then introduced, aiming to apply flexural moments to the joint. The hydraulic jack and lever arm are shown in green and red in Figure 2.21a, respectively. Two pairs of tension-compression load cells (schematically shown in Figure 2.21a-b) are attached to each of the tenon and slot components. The exact location of the load cells in the tenon and slot components is shown in Figure 2.21f and Figure 2.21g, respectively. The vertical movement of the hydraulic jack applies a bending moment to the specimen through the tenon component. The bending moment appears as tension forces on one pair of the load cell and compression forces on the other pair per each tenon or slot component occur. The moment and shear diagrams along the tenon and slot components are shown in Figure 2.21c. Two inclinometers are mounted on each specimen to capture rotations. The measurement devices are attached to the top of the tenon component, as shown in Figure 2.21a. Using two inclinometers was necessary, mainly because the tab insertion angle is 60°, and thus, the geometry of the test specimen is asymmetric. Accordingly, different rotations could occur at the ends of the tenon component. Recalling that the slot component was fully fixed to the steel support and had no rotations during the loading process, the inclinometers recorded the closing angle of the wood-wood connection. The average rotations recorded in the inclinometers are then used to plot the moment-rotation curve.

The average moment, expressed in Eq. 2.2, derived by moving the hydraulic jack, is computed from the multiplication of the lever arm by the uniaxial tension-compression loads in the load cells.

$$M_{avg} = M_{tenon} + M_{slot} = ((a+d).F_{BL} - a.F_{BR}) + ((a+d).F_{TT} - a.F_{TB}) \qquad (2.2)$$

where M_{tenon} and M_{slot} are the flexural moments in the tenon and slot components, respectively. F_{BL} and F_{BR} are the axial forces in the load cells associated with the slot components. F_{TT} and F_{TB} are the axial forces in the load

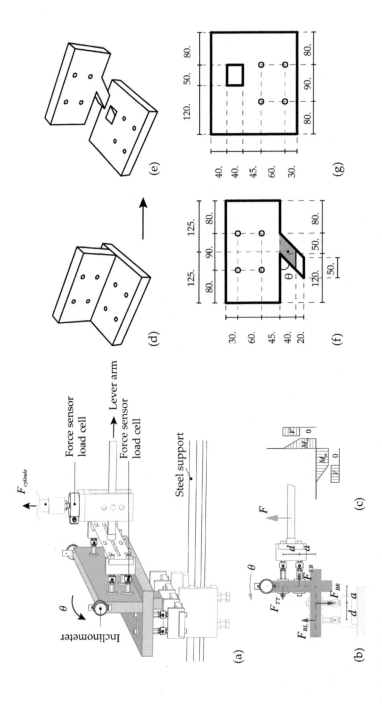

Figure 2.21 The dimension of the specimen and design test setup for the connections under flexural moments.

Note: The geometry of the connections in (a)-(c) is re-generated based on the existing data and the framework investigated and developed by Roche [3].

Credit of images (d)-(g): A. Rezaei Rad, 2020 [49]

cells associated with the tenon components. The terms *a* and *d* are the lever arms between the load cells and the timber components. Figure 2.21b shows the position of the load cells and lever arms with additional detail. In the current test setup, shear forces do not exist in the design specimen, as is shown in Figure 2.21c, indicating that the test setup is efficient and optimized.

2.5.3 Test procedure and loading protocol used for the wood-wood connections

The loading procedure adopted for the tests follows the European protocol for mechanical characterization of joints, known as EN 26891 [1]. Using this protocol, a loading rate relevant to the design parameters of the connections is incorporated. Furthermore, because the protocol requires an unloading-loading step at 40% of the joint capacity, a full interlocking mechanism is established between the tenon and slot components. As such, the effect of fabrication tolerance and associated imperfections on the derived load-deformation or moment-rotation curves would be minimal. Accordingly, the performance of the connection at the serviceability limit state (SLS) and ultimate limit state (ULS) is well understood with a high degree of robustness.

The load rate should be primarily computed to adopt a relevant loading process and calibrate the test setup. To do so, a dummy sample is prepared to estimate the maximum capacity (F_{est}) for each load case. After determining the F_{est}, the test procedure starts with a force-control loading phase applied to the specimen. The state of the force increases only to $0.4F_{est}$, which is then followed by 30 seconds pause. This procedure is mainly done to ensure that a proper interlocking mechanism is established between the tenon and slot components. Next, an unloading process starts, and the load level is decreased to $0.1F_{est}$. The procedure is followed by a 30-second pause, and the load-control phase then continues to augment the load level to $0.7F_{est}$. At this stage, almost all elastic mechanical properties of the joint are determined. To gain insight into the nonlinear, as well as the post-peak behavior of the connection, a secure displacement-control phase is followed by the loading protocol. Generally, the same loading rate used in the first step (the load-control phase) is adopted for this phase. In the wood-wood connections studied herein, a loading rate of 0.1 mm/sec is employed. The displacement-control phase continues until a full collapse occurs in the test specimen. The loading process with respect to time and the different phases involved in the protocol are detailed in Figure 2.22.

2.5.4 Qualitative performance assessment of the wood-wood connections

This section evaluates and discusses the qualitative performance of the through-tenon wood-wood connections as well as investigates the interaction between the tenon and slot components observed during the test

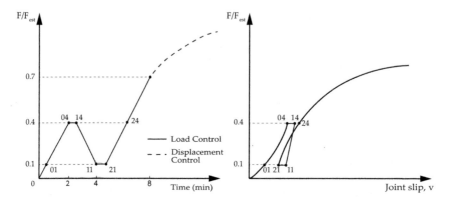

Figure 2.22 **Loading protocol used in the experimental tests and adopted from EN 26891 [1].**

procedure and the load-transferring mechanism for different load cases described earlier. This section aims to support further engineering design by enhancing the detailed understanding of the wood-wood connections, as well as potential failure mechanisms.

The tenon component plays a more important role than the slot component in the load-carrying mechanism for tensile behavior. Given this consideration, the failure in the connection occurs because of a significant loss in the in-between shear resistance of the wood fibers located in the tenon component. The failure surface associated with this type of loss is shown in Figure 2.23. The damage in the tenon components starts with visible cracks at approximately 75% of the total strength. Crack propagation continues until the maximum strength of the connection, where cracks become clearly visible and large. At almost 80% of the connection capacity after the peak force, the connection fails to carry additional tensile loads. The connection failure is relatively brittle; however, strength degradation, especially in the postpeak-force phase, is relatively slow. This is mainly due to the fiber

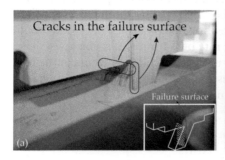

Figure 2.23 **Performance of the wood-wood connections under tensile loads: failure damage state.**

Image credit: A. Rezaei Rad, 2020 [49]

orientation of the tenon and slot components, which provides a favorable postpeak nonlinear behavior. A secondary failure mode is observed in the slot components. In detail, during the loading process, Tenon's tail applies a tensile force to the laminated plywood layers. This, in particular, generates a delamination phenomenon in the slot component. However, cracks are invisible, and the effect of the secondary damage mode is insignificant.

Similar to the tensile behavior, it was observed that the tenon component plays a more important role than the slot component in the load-carrying mechanism for the edgewise behavior. During the experiment, it was perceived that the tenon component undergoes considerable deformations, and consequently, damages. In this load case, the edgewise forces are transmitted from the tenon to the slot component by applying compression forces perpendicular to the fibers of the tenon component. Given that timber fibers behave quite ductile under compression loads, an overall ductile behavior was consequently observed during the loading process. Given the geometry and loads, it is assumed that the tenon component behaves like a simple cantilever beam element subjected to uniformly distributed load. Therefore, the cross section of the tenon component is subjected to bending moments. This indicates that there is a compressive load at the top and a tensile load at the bottom along the tenon length at the cross sections. These tensile and compressive forces are also parallel to the timber fibers.

Given that the tensile capacity of timber is low, the damage in the tenon initiates with multiple cracks at the bottom of the component. As the cracks propagate from the bottom to the top of the cross sections, they grow increasingly larger. Accordingly, they shift the neutral axis from the middle to top until they subject the entire section to the tensile load and cause failure in the tenon component. Figure 2.24 schematically shows the typical

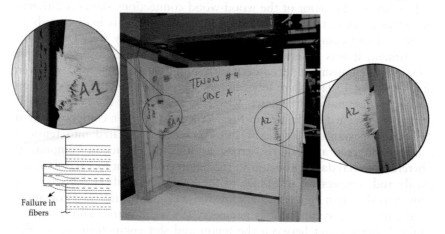

Figure 2.24 Performance of the wood-wood connections under edgewise loads: failure damage state.

Image credit: A. Rezaei Rad, 2020 [49]

failure mode and the damage propagation that occurred in the wood-wood connections under edgewise loads.

Given the configuration of the test setup, the performance of the tenons under the flatwise load was very similar to the edgewise behavior explained in the previous paragraph. In particular, the tenon components behave like an elastic beam mounted on a slot component, and it is subjected to a uniform distributed load. Behaving like an elastic beam element with a certain degree of ductility is directly attributed to the number of longitudinal plywood layers in the tenon. In detail, 11 layers of the cross section are longitudinally oriented in this configuration (Figure 2.16). This dictates that the flatwise load flows from the tenon to the slot component by applying bending moments to the tenon instead of shear loads. During the experiment, it was observed that the bending moments activated several load-resisting mechanisms, especially tension-compression resistance, embedment resistance, shear resistance, and torsional resistance. The contribution of each aforementioned resisting mechanism depends on the fabrication tool and tab insertion angle, and it has been deeply investigated in Rezaei Rad, et al. [9]. In the current investigation, the shear failure in the fibers, the compression failure caused by the interaction between the slot and tenon components, and the tension-compression failure, caused by the bending in the tenon are the main and visible failure mechanisms. Together with a schematic illustration of their position in the tenon component, these failure mechanisms are shown in Figure 2.25. Observing different failure mechanisms and damage types in the connection resulted in a ductile postpeak behavior. Furthermore, having multiple damage indications indicates that the design parameters are well adopted, which leads to a more predictable response than brittle wood-wood connections.

The flexural behavior of the wood-wood connections shows a different performance than the three previous load cases. In this load case, both tenon and slot components play an active and primary role. During the experiment, it was observed that each component gives rise to multiple resisting mechanisms, and consequently, multiple damage modes. Given that 11 out of 13 layers are longitudinally oriented over the cross section of the tenon component (Figure 2.16), the behavior of the connection is primarily governed by the flexural resistance of the tenon component than the shear resistance. The flexural resistance is translated into tension-compression resistance of the cross sections of the tenon component. Therefore, initial damages occur in the tenon component by exceeding the tensile and compressive resistance of the longitudinal fibers. This type of damage is shown in Figure 2.26a. However, because the tab insertion angle is 60°, the tenon component is geometrically asymmetric, and it cannot establish full contact between the tenon and slot component during the entire load-carrying mechanism. As such, the geometry of the connection dictates other resisting mechanisms, especially at higher stages of loads. In particular, torsional resistance and shear resistance were observed as

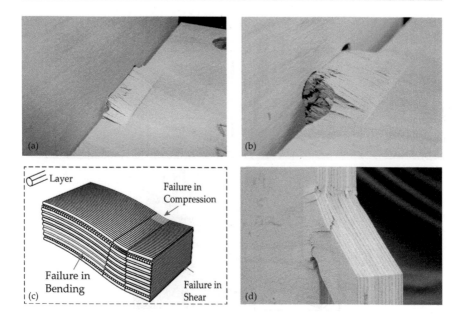

Figure 2.25 Performance of the wood-wood connections under flatwise loads: failure damage state.

Image credit: A. Rezaei Rad, 2020 [49]

the other significant sources of resistance. Although the geometry of the connection is asymmetric, interlocking between the tenon and slot components was fully maintained during the test procedure. This helped the system to optimally transfer the load from the tenon to the slot component. Given this, rotations in the tenon component cause bending moments to the slot component as well. This, in particular, caused the bending failure, as shown in Figure 2.26b, and delamination, as shown in Figure 2.26c.

2.5.5 Constitutive model and quantified performance assessment of the wood-wood connections

The load-deformation behavior of the connections under the tensile, edgewise, and flatwise loads and the moment-rotation behavior of the connections under the flexural moment are shown in Figure 2.27a–d, respectively. The curves indicate that there is a low dispersion among the force-deformation and moment-rotation responses. This, in particular, suggests that the effects of the test setup and replicate-to-replicate variability were minimal. Furthermore, given that the size of the Through-Tenon connections represents the average dimension of such connections in large-scale IATP structures, it is expected that the size effect does not considerably affect the findings.

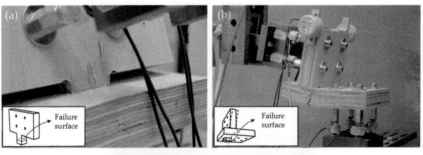

Figure 2.26 **Performance of the wood-wood connections under flexural moments: failure damage state.**

Image credit: A. Rezaei Rad, 2020 [49]

Quantitative evaluation of the observed behavior of the wood-wood connection is provided in this section. The European standards for designing timber elements Eurocode 5 [17] and general principles for determining the strength and deformation characteristics recommended by EN 26891 [1] are used. The performance measures are summarized in the maximum strength and design stiffness – also known as slip modulus. Summarized in Table 2.1, these performance measures are selected because the design space is limited to linear elastic analysis in spatial timber plate structures. Other design parameters such as ductility, yield, and ultimate strength, and associated deformation/rotation can be potentially considered for nonlinear performance assessment.

2.6 KINEMATIC OF TIMBER PLATES EQUIPPED WITH WOOD-WOOD CONNECTIONS[5]

After gaining insight into the behavior of the wood-wood connections, the primary focus is put on determining the behavior of timber plate elements with wood-wood connections. Toward this goal, the main

---

[5] This section uses the material published in A. Rezaei Rad, et al. [18] under the terms of the Creative Commons Attribution 4.0 International license (CC BY 4.0).

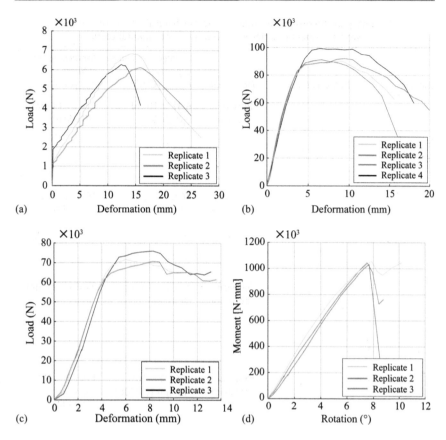

Figure 2.27 The load-deformation curve corresponding to the (a) tensile load, (b) edge-wise load, (c) flatwise load, and (d) moment-rotation curve corresponding to the flexural moment.

Research credit: A. Rezaei Rad, 2020 [49]

kinematics associated with an IATP component is first recognized. The term *component* represents a timber plate and the wood-wood connections that are located around its edge. This methodology suggests that the behavior of a timber component is summarized in the in-plane and out-of-plane response. In this section, the in-plane and out-of-plane load-carrying mechanisms of an IATP component and the equations governing these mechanisms are provided.

To develop the governing equations, only the relative displacements associated with the perimeter of the plate are considered for both the in-plane and out-of-plane kinematics. In other words, only the degrees of freedom (DOF) associated with nodes located at the perimeter of the component are used. This assumption significantly simplifies the process and reduces the size of the system of equations to be solved. Consequently, the

Table 2.1 Summary of the design parameters and mechanical properties of the wood-wood connections derived from the experimental tests

Kinematic	Representation	Description	Average maximum capacity $F_{max,avg}$ [COV]	Average design stiffness K_{avg} [COV]
		Tensile behavior	6.47×10^3 N [0.0532]	416.81 N/mm [0.0616]
		Edgewise behavior	46.83×10^3 N [0.0448]	15009.24 N/mm [0.0179]
		Flatwise behavior	36.55×10^3 N [0.0388]	9489.04 N/mm [0.0274]
		Flexural behavior	1067.0×10^3 N.mm [0.0708]	170.19×10^3 N.mm/° [0.0484]

computational cost is lowered, and a relatively straightforward numerical model can be used to simulate the behavior. The developed equations are then used to develop detailed numerical models, which will be widely discussed and presented in later sections.

2.6.1 In-plane behavior of IATP components

2.6.1.1 Recognized degrees of freedom

To study the in-plane behavior of IATP components, the load cases which can potentially exist in freeform geometries are considered herein. This ensures that the subsequent governing equations will also be valid for IATP elements used in simpler geometries. Given this, an IATP component can be subjected to the tension, compression, and shear in-plane forces and in-plane flexural moment. The role of each load case in IATP structures, the interaction between different timber plates under each load case, and the associated deformations that occur in timber plates are described in the following paragraphs.

- In the first load case shown in Figure 2.28a, an adjacent IATP component applies tension forces to the element of interest. For example, the timber plates within Box #2 and #3 in Figure 2.5 are adjacent to the T_1 and B_1 plates of Box #1. In this case, Boxes #2 and #3 can apply tension forces to these plates. The forces are shown as P_X and P_Z in Figure 2.28a. The in-plane tension loads are transferred from the through-tenon joints indexed 1 to 5 to the T_1 and B_1 plates and exited the through-tenon joints indexed 1' to 5'. The undeformed and deformed T_1 and B_1 are shown in Figure 2.28b.
- In the second load case shown in Figure 2.28b, an adjacent IATP component applies compression forces to the element of interest. In this case, because of the edgewise contact between the elements, the entire perimeter of the component carries the compression force and transfers the load to its neighbor. For instance, and recalling Figure 2.5, the plate CT_2 in Box #2 and the plate CL_3 in Box #3 are in edgewise contact with the plates T_1 and B_1 of Box #1, and they can apply compression forces to these elements. The forces are shown as P_X and P_Z in Figure 2.28a. Furthermore, similar to the first load case, Figure 2.28b shows the undeformed and deformed shapes of the plates T_1 and B_1.
- In the third load case shown in Figure 2.28c, an adjacent IATP component applies lateral forces to the element of interest. This force can result in shear behavior in the element of interest. For instance, and recalling Figure 2.5, the plate CT_2 in Box #2 and the plate CL_3 in Box #3 can apply lateral force to T1 and B1 plates in Box #1. The force associated with the shear behavior in Box #1 is shown as P_{XZ} in Figure 2.28a. The lateral force is transferred from the through-tenon joints indexed 1 to 5 to the plates T_1 and B_1 and exited the through-tenon joints indexed 1' to 5'. The undeformed and deformed plates T_1 and B_1 are shown in Figure 2.28b.
- In the fourth load case shown in Figure 2.28d, an adjacent IATP component applies flexural moments to the element of interest. For instance, consider the out-of-plane loads applied to the plates T_1 and B_1 in Box #1, shown as P_Y, P_Z, and M_X in Figure 2.28c. These loads are transmitted via the through-tenon joints, and they impose in-plane flexural moments on the plates CL_1 and CT_1 in Box #1. The in-plane flexural moment, shown as M_X, is applied to the plates at the through-tenon joints indexed 1 to 6. The moment and the undeformed and deformed shapes of the plate CL_1 are shown in Figure 2.28d. As another example, the plate CT_2 of Box #2 and the plate CL_3 of Box #3 can apply flexural moments to the plates T_1 and B_1 in Box #1. This flexural moment is shown as M_Y in Figure 2.28a-b.

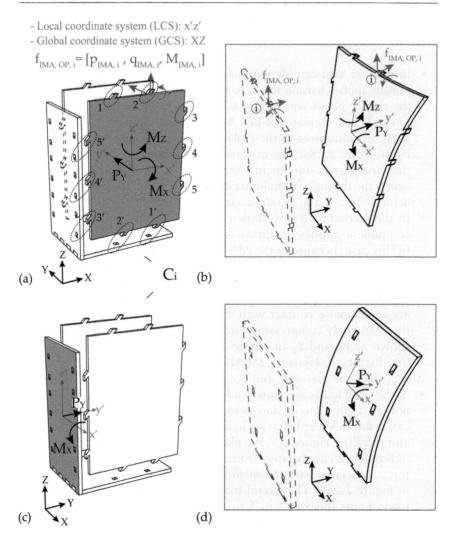

- Local coordinate system (LCS): x'z'
- Global coordinate system (GCS): XZ

$$f_{IMA,\,OP,\,i} = [p_{IMA,\,i}\,,\,q_{IMA,\,i}\,,\,M_{IMA,\,i}]$$

(a)   (b)   (c)   (d) C_i

* In the undeformed/undisplaced condition, The LCS and GCS are similar.

Figure 2.28 In-plane loadin g condition, undeformed and deformed shapes for the IATP element (a, b) top/bottom plate and (c,d) cross longitudinal plates.

Figure credit: A. Rezaei Rad, et al. [18].

2.6.1.2 Forces and deformations

The equations governing the in-plane behavior of the T_i and B_i plates, the associated local and global coordinate systems, and the applied loads shown in Figure 2.28a-b, are formulated in this section. Five kinematic DOFs, including uniform shear (γ_{XZ}, Figure 2.29a) and uniform flexural deformations (ϕ, Figure 2.29b), two rigid body translations

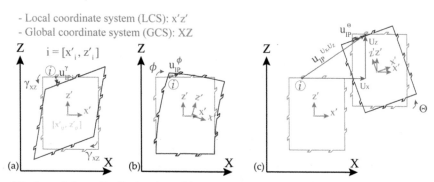

- Local coordinate system (LCS): x'z'
- Global coordinate system (GCS): XZ

* In the undeformed/undisplaced condition, The LCS and GCS are similar.
* Local x'z' plane is always parallel to global XZ plane.

Figure 2.29 Sources of in-plane deformations for an IATP component including (a) shear, (b) flexural deformation, and (c) rigid body translation and rotation.

Figure credit: A. Rezaei Rad, et al. [18]. A. Rezaei Rad, 2020 [49]

(U_X and U_Y, Figure 2.29c), and rigid body rotation (Θ, Figure 2.29c), are defined for each IATP element. The vector D_{IP}, which is defined as $D_{IP} = \{U_X \quad U_Z \quad \gamma_{XZ} \quad \Theta \quad \phi\}^T$, is used to describe these kinematic DOFs in the global coordinate system (GCS) (X, Z) when the element responds within the linear elastic range.

The force vector $F_{IP} = \{P_X \quad P_Z \quad P_{XZ} \quad M_{Y,rgd} \quad M_{Y,flx}\}^T$, which corresponds to the vector D_{IP}, is defined, where P_X and P_Y are the in-plane forces along the global X and Z axes, respectively, and P_{XZ} is the in-plane shear force. The in-plane moment M_Y, shown in Figure 2.28b, is decomposed into two components: $M_{Y,rgd}$, which comes from the neighboring boxes and causes rigid body rotations, and $M_{Y,flx}$, which is the in-plane moment that causes flexural deformations.

The in-plane displacement of node i, which represents the ith connection located at the perimeter of the timber plate (Figure 2.29), is described by the vector $u_{IP,i} = \{X_{IP,i} \quad Z_{IP,i}\}^T$ in the GCS (X, Z). The vector includes contributions from shear deformation $u_{IP,i}^{\gamma_{XZ}}$, flexural deformation $u_{IP,i}^{\phi}$, rigid body translation $u_{IP,i}^{U_X,U_Z}$, and rigid body rotation $u_{IP,i}^{\Theta}$. The superposition principle is used to compute $u_{IP,i}$ (Eq. 2.3), which is deemed valid because only linear elastic responses are considered.

$$u_{IP,i} = u_{IP,i}^{U_X,U_Z} + u_{IP,i}^{\gamma_{XZ}} + u_{IP,i}^{\phi} + u_{IP,i}^{\Theta} \tag{2.3}$$

The components of the vector $u_{IP,i}$, which correspond to the in-plane displacement field of the IATP element, were described in the GCS in the previous paragraph. However, in the next section, it will be shown that to compute the element stiffness, it is necessary to express the components of $u_{IP,i}$ in the local coordinate system (LCS) (x',z'). Furthermore, the definition

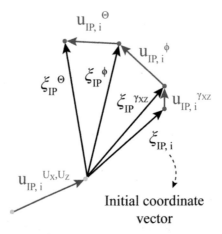

$$u_{IP,i}^{\Theta}$$
$$u_{IP,i}^{\phi}$$
$$\xi_{IP}^{\Theta} \quad \xi_{IP}^{\phi}$$
$$\xi_{IP}^{\gamma xz} \quad u_{IP,i}^{\gamma xz}$$
$$\xi_{IP,i}$$
$$u_{IP,i}^{U_X,U_Z}$$

Initial coordinate
vector

Figure 2.30 Coordinate and deformation vector relationship used to express the components of **u**$_{IP,i}$ in the local coordinate system.

Figure credit: A. Rezaei Rad, et al. [18]. A. Rezaei Rad, 2020 [49]

of the LCS is necessary to determine the location of node i at each load/displacement step. To define the LCS, a coordinate vector $\xi_{IP,i} = \{x'_i \quad z'_i\}^T$ is defined. This vector specifies the position of node i in the LCS with respect to its origin at each step. The origin of the LCS is located at the center of the panel (Figure 2.29), and in the undeformed/undisplaced condition, is assumed to be parallel to the GCS. As noted earlier, the vector $\mathbf{u}_{IP,i}$ is computed as the vector sum of the deformations/displacements from the various sources (Eq. 2.3). Consequently, $\xi_{IP,i}$ is updated at each step and used for the calculations in the next step.

To express the components of $\mathbf{u}_{IP,i}$ in the local coordinate system, the consecutive deformations/displacements, including rigid body translations, shear deformations, flexural deformations, and rigid body rotations are first recalled. Figure 2.30 shows the vector relationship between $\mathbf{u}_{IP,i}$ and $\xi_{IP,i}$. The coordinate vectors, corresponding to the recognized displacement/deformation fields, are those associated with shear deformation $\xi_{IP,i}^{\gamma xz}$, flexural deformation $\xi_{IP,i}^{\phi}$, and rigid body rotation $\xi_{IP,i}^{\Theta}$.

Recalling that the LCS of an IATP element changes at each step, the coordinate vector, $\xi_{IP,i}$, should be updated. It is worth noting that the rigid body translations do not cause any translation nor rotation in the LCS, and therefore, the coordinate vector remains unchanged after they are applied. Recalling Eq. 2.3, the displacement vector is re-expressed in Eq. 2.4 in terms of the recognized in-plane DOFs and the local coordinate vectors.

$$\mathbf{u}_{IP,i} = \begin{Bmatrix} U_X \\ U_Z \end{Bmatrix} + \left[\xi_{IP,i}^{\Theta} - \xi_{IP,i}^{\phi} \right] + \left[\xi_{IP,i}^{\phi} - \xi_{IP,i}^{\gamma xz} \right] + u_{IP,i}^{\gamma xz} \qquad (2.4)$$

Where the term $\left\{ \begin{array}{c} U_X \\ U_Z \end{array} \right\}$ describes the rigid body translation ($u_{IP,i}^{U_X,U_Z}$ in Eq. 2.3), $\left[\xi_{IP,i}^{\theta} - \xi_{IP,i}^{\phi} \right]$ is associated with the rigid body rotation ($u_{IP,i}^{\theta}$ in Eq. 2.3), $\left[\xi_{IP,i}^{\phi} - \xi_{IP,i}^{\gamma xz} \right]$ is related to the uniform flexural deformation ($u_{IP,i}^{\phi}$ in Eq. 2.3), and $u_{IP,i}^{\gamma xz} = \gamma.\{ z'\ x' \}^T$ describes the shear deformation ($u_{IP,i}^{\gamma xz}$ in Eq. 2.3). Assuming a uniform in-plane flexural deformation, it follows that $\xi_{IP,i}^{\theta} = \Lambda(\Theta).\xi_{IP,i}^{\phi}$. Therefore, Eq. 2.4 is re-formulated in the form of Eq. 2.5.

$$u_{IP,i} = \left\{ \begin{array}{c} U_X \\ U_Z \end{array} \right\} + \left[\Lambda(\Theta).\xi_{IP,i}^{\phi} - \xi_{IP,i}^{\phi} \right] + \left[\xi_{IP,i}^{\phi} - \xi_{IP,i}^{\gamma xz} \right] + u_{IP,i}^{\gamma xz} \qquad (2.5a)$$

$$u_{IP,i} = \left\{ \begin{array}{c} U_X \\ U_Z \end{array} \right\} + \Lambda(\Theta).\xi_{IP,i}^{\phi} - \xi_{IP,i}^{\gamma xz} + u_{IP,i}^{\gamma xz} \qquad (2.5b)$$

Where $\Lambda(\Theta) = \begin{bmatrix} \cos(\Theta) & -\sin(\Theta) \\ \sin(\Theta) & \cos(\Theta) \end{bmatrix}$ is the transformation matrix for rigid body rotation. Referring to Figure 2.30, it follows that $\xi_{IP,\,i}^{\phi} = \Lambda(\phi).\xi_{IP,\,i}^{\gamma xz}$, and therefore, Eq. 2.5b is re-written in the form of Eq. 2.6.

$$u_{IP,\,i} = \left\{ \begin{array}{c} U_X \\ U_Z \end{array} \right\} + \Lambda(\phi).\Lambda(\Theta).\xi_{IP,\,i}^{\gamma xz} - \xi_{IP,\,i}^{\gamma xz} + u_{IP,\,i}^{\gamma xz} \qquad (2.6)$$

Where $\Lambda(\phi) = \begin{bmatrix} \cos(\phi) & -\sin(\phi) \\ \sin(\phi) & \cos(\phi) \end{bmatrix}$ is the transformation matrix for in-plane flexural deformations. Referring to Figure 2.30, it follows that $\xi_{IP,i}^{\gamma xz} = \xi_{IP,i} + u_{IP,i}^{\gamma xz}$, and therefore, Eq. 2.6 is re-written in the form of Eq. 2.7.

$$u_{IP,\,i} = \left\{ \begin{array}{c} U_X \\ U_Z \end{array} \right\} + \Lambda(\phi).\Lambda(\Theta).\left(\xi_{IP,\,i} + u_{IP,\,i}^{\gamma xz} \right) - \left(\xi_{IP,\,i} + u_{IP,\,i}^{\gamma xz} \right) + u_{IP,\,i}^{\gamma xz} \qquad (2.7)$$

Given that $\xi_{IP,i} = \{ x'_i\ z'_i \}^T$ and $u_{IP,i}^{\gamma xz} = \gamma . \{ z'\ x' \}^T$, Eq. 2.7 is written in the form of Eq. 2.8.

$$u_{IP,\,i} = \left\{ \begin{array}{c} U_X \\ U_Z \end{array} \right\} + \Lambda(\phi).\Lambda(\Theta).\left(\left\{ x'_i\ z'_i \right\}^T + \gamma.\left\{ z'\ x' \right\}^T \right) - \left\{ x'_i\ z'_i \right\}^T \qquad (2.8)$$

Applying $\Lambda(\Theta) = \begin{bmatrix} \cos(\Theta) & -\sin(\Theta) \\ \sin(\Theta) & \cos(\Theta) \end{bmatrix}$ and $\Lambda(f) = \begin{bmatrix} \cos(\phi) & -\sin(\phi) \\ \sin(\phi) & \cos(\phi) \end{bmatrix}$, Eq. 2.8 is written in the form of Eq. 2.9.

$$u_{IP,i} = \left\{ \begin{array}{c} U_X \\ U_Z \end{array} \right\} + \left[\begin{bmatrix} \cos(\phi) & -\sin(\phi) \\ \sin(\phi) & \cos(\phi) \end{bmatrix} \cdot \begin{bmatrix} \cos(\Theta) & -\sin(\Theta) \\ \sin(\Theta) & \cos(\Theta) \end{bmatrix} \right.$$

$$\left. \cdot \left(\left\{ \begin{array}{c} x' \\ z' \end{array} \right\} + \gamma \cdot \left\{ \begin{array}{c} z' \\ x' \end{array} \right\} \right) \right] - \left\{ \begin{array}{c} x' \\ z' \end{array} \right\} \qquad (2.9a)$$

$$u_{IP,i} = \left\{ \begin{array}{c} U_X \\ U_Z \end{array} \right\} + \left[\begin{bmatrix} \cos(\Theta).\cos(\phi) - \sin(\Theta)\sin(\phi) & -\cos(\Theta).\sin(\phi) - \sin(\Theta)\cos(\phi) \\ \sin(\Theta).\cos(\phi) + \cos(\Theta)\sin(\phi) & -\sin(\Theta)\sin(\phi) + \cos(\Theta)\cos(\phi) \end{bmatrix} \right.$$

$$\left. \cdot \left\{ \begin{array}{c} x' + \gamma.z' \\ z' + \gamma.x' \end{array} \right\} \right] - \left\{ \begin{array}{c} x' \\ z' \end{array} \right\} \qquad (2.9b)$$

Finally, the total deformation of an IATP element in a GCS can be expressed in terms of LCS indicators in Eq. 2.10.

$$u_{IP,i} = \left\{ \begin{array}{c} X_{IP,i} \\ Z_{IP,i} \end{array} \right\} = \left\{ \begin{array}{c} U_X \\ U_Z \end{array} \right\}$$

$$+ \left[\begin{array}{c} (\cos(\Theta).\cos(\phi) - \sin(\Theta)\sin(\phi))(x' + \gamma.z') - (\cos(\Theta).\sin(\phi) + \sin(\Theta)\cos(\phi))(z' + \gamma.x') \\ (\sin(\Theta).\cos(\phi) + \cos(\Theta)\sin(\phi))(x' + \gamma.z') - (\sin(\Theta)\sin(\phi) - \cos(\Theta)\cos(\phi))(z' + \gamma.x') \end{array} \right]$$

$$- \left\{ \begin{array}{c} x' \\ z' \end{array} \right\} \qquad (2.10)$$

2.6.1.3 Virtual work principle

So far, the displacement and the force fields corresponding to the in-plane behavior of IATP components are determined. To compute the in-plane stiffness of an IATP element, the virtual work principle, which is expressed in Eq. 2.11, is applied.

$$\delta W_{external} = \delta W_{internal} \qquad (2.11)$$

The external work is done by the external force, F_{IP} on the virtual displacement δD_{IP}^T, while the internal work is done by the compatible virtual deformations/strains ($\delta \gamma_x$ and $\delta \varepsilon_x$ for timber plates and $\delta u_{IP,j}$ for connections) on the element forces/stresses (the axial stress $[\sigma_x]$ and shear stress $[\tau_x]$ for the timber plate, and $f_{IMA,IP,j}$ for the connections). It is worth

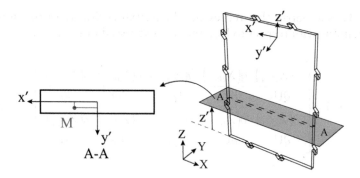

Figure 2.31 IATP element and a cross-section used for the in-plane formulation with the virtual work principle.

Figure credit: A. Rezaei Rad, et al. [18].

noting that γ_x and ε_x are shear and axial strains at point M in Figure 2.31. Accordingly, τ_x and σ_x are the shear and axial stresses. Eq. 2.11 is then rewritten in the form of Eq. 2.12.

$$\delta D_{IP}^{T}.F_{IP} = \sum_{j=1}^{m}\left[\delta u_{IP,j}{}^{T}.f_{IMA,IP,j}\right] + \int_{V}\delta\varepsilon^{T}.\sigma dV \tag{2.12a}$$

$$\delta D_{IP}^{T}.F_{IP} = \sum_{j=1}^{m}\left[\delta u_{IP,j}{}^{T}.f_{IMA,IP,j}\right] + \int_{L}\int_{A}\delta\gamma_{z'}{}^{T}.\tau_{z'}dAdz' + \int_{L}\int_{A}\delta\varepsilon_{z'}{}^{T}.\sigma_{z'}dAdz' \tag{2.12b}$$

where the term $\Sigma_{j=1}^{m}[\delta u_{IP,j}{}^{T}.f_{IMA,IP,j}]$ determines the contribution of the wood-wood connections, $\int_{L}\int_{A}\delta\gamma_{z'}{}^{T}.\tau_{z'}dAdz'$ determines the shear contribution of the plate, and $\int_{L}\int_{A}\delta\varepsilon_{z'}{}^{T}.\sigma_{z'}dAdz'$ determines the axial-flexural contribution of the plate to the in-plane response. In detail, F_{IP} and δD_{IP}^{T} are the external force and external virtual displacement, respectively. $\delta\varepsilon_{z'}$ and $\delta\gamma_{z'}$ correspond to the virtual internal axial and shear strains for the timber plate at point M in Figure 2.31, respectively. $\sigma_{z'}$ and $\tau_{z'}$ are the internal axial and shear stresses for the timber plate, respectively. According to Figure 2.31, the strains and stresses are the function of the position z' along the element axis and the position of point M within the cross section [19].

Given the first term in the right hand of Eq.2.12b, the term $f_{IMA,IP,j}$ is the internal force for the connections, and $\delta u_{IP,j}$ is the virtual internal displacement for the wood-wood connections. The former, which is shown in Figure 2.28a-b, is defined as $f_{IMA,IP,i} = \{p_{IMA,IP,i} \, q_{IMA,IP,i}\}^{T}$, and the latter corresponds to the virtual form of $u_{IP,j}$, and it is expressed in Eq. 2.13.

$$\delta u_{IP,i} = B_{IP}.\delta D_{IP} \tag{2.13}$$

where B_{IP} is a matrix that represents the partial derivative of the in-plane deformations with respect to D_{IP}, and it is expressed in Eq. 2.14.

$$B_{IP}(D_{IP}) = \begin{bmatrix} \dfrac{\partial(X_{IP,\,i})}{\partial U_X} & \dfrac{\partial(X_{IP,\,i})}{\partial U_Z} & \dfrac{\partial(X_{IP,\,i})}{\partial \gamma_{XZ}} & \dfrac{\partial(X_{IP,\,i})}{\partial \Theta} & \dfrac{\partial(X_{IP,\,i})}{\partial \phi} \\[3mm] \dfrac{\partial(Z_{IP,\,i})}{\partial U_X} & \dfrac{\partial(Z_{IP,\,i})}{\partial U_Z} & \dfrac{\partial(Z_{IP,\,i})}{\partial \gamma_{XZ}} & \dfrac{\partial(Z_{IP,\,i})}{\partial \Theta} & \dfrac{\partial(Z_{IP,\,i})}{\partial \phi} \end{bmatrix} \qquad (2.14)$$

Where,

$$\frac{\partial(X_{IP,\,i})}{\partial U_X} = \frac{\partial(Z_{IP,\,i})}{\partial U_Z} = 1$$

$$\frac{\partial(X_{IP,\,i})}{\partial U_Z} = \frac{\partial(Z_{IP,\,i})}{\partial U_X} = 0$$

$$\frac{\partial(X_{IP,i})}{\partial \Theta} = \left(-\sin(\Theta)\cos(\phi) - \cos(\Theta)\sin(\phi)\right)(x' + \gamma.z')$$
$$+ \left(\sin(\Theta).\sin(\phi) - \cos(\Theta)\cos(\phi)\right)(z' + \gamma.x')$$

$$\frac{\partial(X_{IP,i})}{\partial \phi} = \left(-\sin(\phi).\cos(\Theta) - \cos(\phi).\sin(\Theta)\right)(x' + \gamma.z')$$
$$+ \left(-\cos(\Theta)\cos(\phi) + \sin(\Theta).\sin(\phi)\right)(z' + \gamma.x')$$

$$\frac{\partial(X_{IP,i})}{\partial \gamma_{XZ}} = \left(\cos(\Theta).\cos(\phi) - \sin(\Theta)\sin(\phi)\right).z'$$
$$+ \left(-\cos(\Theta).\sin(\phi) - \sin(\Theta)\cos(\phi)\right).x'$$

$$\frac{\partial(Z_{IP,i})}{\partial \Theta} = \left(\cos(\Theta).\cos(\phi) - sin(\Theta).\sin(\phi)\right)(x' + \gamma.z')$$
$$+ \left(-\cos(\Theta).\sin(\phi) - \sin(\Theta)\cos(\phi)\right)(z' + \gamma.x')$$

$$\frac{\partial(Z_{IP,i})}{\partial \phi} = \left(-\sin(\phi).\sin(\Theta) + \cos(\phi).\cos(\Theta)\right)(x' + \gamma.z')$$
$$+ \left(-\sin(\Theta)\cos(\phi) - \cos(\Theta).sin(\phi)\right)(z' + \gamma.x')$$

$$\frac{\partial\left(Z_{IP,i}\right)}{\partial\gamma_{XZ}} = \left(\sin(\Theta).\cos(\phi) + \cos(\Theta)\sin(\phi)\right).z'$$

$$+\left(-\sin(\Theta)\sin(\phi) + \cos(\Theta)\cos(\phi)\right).x'$$

Given the second term in the right hand of Eq. 2.11b, the strains and stresses are a function of the position x' along the element axis and the position of point M (Figure 2.31) within the cross section [19]. Regarding the shear contribution, the shear strain at point M in Figure 2.31, $\gamma_{z'}(x',y',z')$, is written as a function of section deformation, $\gamma(z')$ and strain distribution, $a_s(y',x')$ in Eq. 2.15.

$$\gamma_{z'}(x',y',z') = a_s(y',x').\gamma(z') \tag{2.15}$$

A uniform distribution of strain and deformations along the profile is assumed [20], which results in $\gamma_{z'}(x',y',z') = \gamma$, where γ has been already defined in Figure 2.29.

Given the third term in the right hand of Eq. 2.12b, the axial strain at point M in Figure 2.31, $\varepsilon_{z'}(x',y',z')$, is written as a function of section deformation, $e(z')$, and strain distribution, $a_s(y',x')$ in Eq. 2.16.

$$\varepsilon_x(x,y,z) = a_s(y,z).e(x) \tag{2.16}$$

Given that Bernoulli's assumption is valid for the in-plane flexural deformations, the axial strain distribution function results in $a_s(y',x') = [1 \quad -y' \quad x']$ [19]. As far as the in-plane behavior is concerned, the rotation about the local x' axis is ignored and accordingly $a_s(y',x') = [1 \quad -y']$. Therefore, Eq. 2.12b is re-written in the form of Eq. 2.17.

$$\delta D_{IP}^T.F_{IP} = \sum_{j=1}^{m}\left[\delta D_{IP}^T.B_{IP}^T.f_{IMA,IP,j}\right] + \int_L \delta\gamma^T \int_A \tau(x)dAdx$$

$$+ \int_L \delta e(x)^T \int_A \begin{pmatrix} 1 \\ -y \end{pmatrix}\sigma(x)dAdx \tag{2.17}$$

To simplify Eq. 2.17, the terms $S_\gamma(z') = \int_A \tau(z')dA$ and $S_\varepsilon(z') = \int_A \begin{pmatrix} 1 \\ -y' \end{pmatrix}\sigma(z')dA$ are defined as the section forces for the shear and axial virtual works, respectively. Given that the stress is uniformly distributed, the section forces are such that $S_\gamma(z') = V(z')$ and $S_\varepsilon(z') = \begin{pmatrix} N(z') \\ M_z(z') \end{pmatrix}$.

The terms $V(z')$, $N(z')$, and $M_z(z')$ are the shear and axial forces and bending moment about the local y' axis in section A-A (Figure 2.31).

2.6.1.4 In-plane stiffness

Assuming that there are 'm' number of wood-wood connections along the edge of the timber plate, the virtual work principle is applied, and Eq. 2.17 is recalled. Eliminating the virtual terms, the applied loads are expressed in terms of internal forces in the matrix form in Eq. 2.18.

$$F_{IP} = \sum_{j=1}^{m} \left[(B_{IP})^T \cdot f_{IMA,IP,j} \right] + \begin{Bmatrix} \int_L N(z')dx \\ 0 \\ \int_L V(z')dx \\ 0 \\ \int_L M_{y'}(z')dx \end{Bmatrix} \tag{2.18}$$

Where F_{IP} is the global external force vector, B_{IP} is the partial derivative matrix of the in-plane deformation with respect to D_{IP}, $f_{IMA,IP,j}$ is the internal force vector of each wood-wood connection, $\int_L N(z')dx$, $\int_L V(z')dx$, and $\int_L M_{y'}(z')dx$ are the internal axial force in the local z' direction, the in-plane shear force and the bending moment of the timber plate about local y', respectively.

Finally, given the external force (F_{IP}) and displacement (D_{IP}) vectors, the in-plane stiffness of the IATP component is defined as $\frac{\partial F_{IP}}{\partial D_{IP}}$ ratio and expressed in Eq. 2.19.

$$K_{\text{in-plane, IATP}} = \frac{\partial F_{IP}}{\partial D_{IP}} = \frac{\partial}{\partial D_{IP}} \left(\sum_{j=1}^{m} \left[B_{IP,j}{}^T \cdot f_{IMA,IP,j} \right] + \begin{Bmatrix} \int_L N(z')dz' \\ 0 \\ \int_L V(z')dz' \\ 0 \\ \int_L M_z(z')dz' \end{Bmatrix} \right) \tag{2.19}$$

Expanding Eq. 2.19, it follows that:

$$K_{\text{In-plane, IATP}} = \sum_{j=1}^{m}\left(B_{\text{IP},j}{}^{T} \cdot \frac{\partial f_{IMA,IP,j}}{\partial D_{\text{IP}}} \right) + \sum_{j=1}^{m}\left(f_{IMA,IP,j} \cdot \frac{\partial B_{\text{IP},j}{}^{T}}{\partial D_{\text{IP}}} \right) + \frac{\partial}{\partial D_{\text{IP}}}\left\{ \begin{array}{c} \int_{L} N(z')dz' \\ 0 \\ \int_{L} V(z')dz' \\ 0 \\ \int_{L} M_{z}(z')dz' \end{array} \right\}$$

(2.20)

$$K_{\text{In-plane, IATP}} = \sum_{j=1}^{m}\left(B_{\text{IP},j}{}^{T} \cdot \frac{\partial f_{IMA,IP,j}}{\partial u_{\text{IP},i}} \cdot \frac{\partial u_{\text{IP},i}}{\partial D_{\text{IP}}} \right) + \sum_{j=1}^{m}\left(f_{IMA,IP,j} \cdot \frac{\partial B_{\text{IP},j}{}^{T}}{\partial D_{\text{IP}}} \right) + \frac{\partial}{\partial D_{\text{IP}}}\left\{ \begin{array}{c} \int_{L} N(z')dz' \\ 0 \\ \int_{L} V(z')dz' \\ 0 \\ \int_{L} M_{z}(z')dz' \end{array} \right\}$$

(2.21)

The higher-order effects, which are reflected in $\sum_{j=1}^{m}\left(f_{IMA,IP,\,j} \cdot \frac{\partial B_{\text{IP},j}{}^{T}}{\partial D_{\text{IP}}} \right)$, are neglected, and the in-plane stiffness of an IATP element, including the plate (K_{Plate}) and the IMAs ($K_{\text{Connection}}$), is written in the form of Eq. 2.22 and Eq. 2.23.

$$K_{\text{In-plane, IATP}} = K_{\text{plate}} + K_{\text{connection}} = \sum_{j=1}^{m}\left(B_{\text{IP},j}{}^{T} \cdot K_{IMA} \cdot B_{\text{IP},j} \right) + K_{\text{Plate}}$$

(2.22)

$$K_{\text{In-plane, IATP}} = \sum_{j=1}^{m}\left(B_{\text{IP},j}{}^{T} \cdot \begin{bmatrix} K_{IMA,tns} & 0 \\ 0 & K_{IMA,edg} \end{bmatrix} \cdot B_{\text{IP},j} \right)$$

$$+ \begin{bmatrix} K_{Plate.ax,0} & 0 & 0 & 0 & 0 \\ & 0 & 0 & 0 & 0 \\ & & K_{Plate.Shr} & 0 & 0 \\ & sym. & & 0 & 0 \\ & & & & K_{Plate.flx} \end{bmatrix}$$

(2.23)

Where $K_{Plate.ax,0}$ is the axial stiffness along the fiber-parallel direction, $K_{Plate.Shr}$ is the in-plane shear stiffness associated with the relative displacement along the local z' axis, and $K_{Plate.flx}$ is the flexural stiffness about the local y'.

2.6.2 Out-of-plane behavior of IATP components

2.6.2.1 Recognized degrees of freedom

To study the out-of-plane behavior of IATP elements, the load cases that can potentially exist in IATP structures with free-form geometries and give rise to the out-of-plane behavior of IATP components are considered in the formulation. Given this consideration, an IATP element can be subjected to two main out-of-plane load cases. These load cases, which are elaborated in the next two paragraphs, mainly depend on the distribution of wood-wood connections along the perimeter of the timber plate. Furthermore, the loads applied by the neighboring elements on the element of interest and any external load applied directly to the element are considered in the formulation.

- In the first load case, the element of interest is subjected to biaxial bending. Given that an IATP element is quadrilateral, the biaxial load case occurs when two neighboring edges of the element are constrained while the other two edges are subjected to out-of-plane loads. For instance, and considering Figure 2.32a, the T_8/B_8 plates are subjected to out-of-plane loads by the neighboring elements via the connections joints indexed 1 to 5. These joints are located at two edges of the plate. On the other hand, the element is supported by the neighboring elements via the connections joints indexed 1' to 5', which are located at the other two edges of the plate. In other words, these two edges are constrained. Figure 2.32b shows the loads applied to the T_8/B_8 plates (P_Y, M_X, and M_Z), as well as the undeformed and deformed shapes.
- In the first load case, the element of interest is subjected to uniaxial bending. This load case mainly occurs when only one edge of the quadrilateral is constrained (i.e., supported by the neighboring element), while the other edges are subjected to out-of-plane loads. For instance, and considering Figure 2.32c, the CL_8 plate is supported by the dovetail joints, which are located at the bottom of the plate. As such, only one edge of the plate is constrained and supports the plate. On the other hand, the connections that apply out-of-plane loads, are located at the other edges of the plate. Figure 2.32d shows the out-of-plane loads (P_Y, M_x) applied to the CL_8 plate, as well as the undeformed and deformed shapes under uniaxial bending.

Given that the plate thickness is small relative to the perimeter dimensions, the out-of-plane behavior of IATPs is characterized using multiple

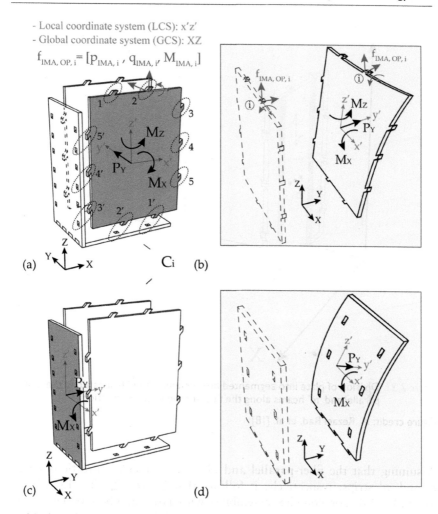

- Local coordinate system (LCS): x'z'
- Global coordinate system (GCS): XZ

$$f_{IMA, OP, i} = [p_{IMA, i}, q_{IMA, i}, M_{IMA, i}]$$

C_i

(a)   (b)   (c)   (d)

* In the undeformed/undisplaced condition, The LCS and GCS are similar.

Figure 2.32 The Out-of-plane loading condition and undeformed/deformed shapes for (a, b) Top/Bottom plates (biaxial bending about the local x' and z' axes), (c, d) Cross plates (uniaxial bending about the local x' axis).

Figure credit: A. Rezaei Rad, et al. [18].

beam elements (i.e., to capture biaxial bending). Figure 2.33 shows the beam idealization of an IATP element. The beam elements, which represent a strip of the plate, are along the direction parallel and perpendicular to the fibers. The contribution to out-of-plane bidirectional bending from each fiber-parallel and fiber-perpendicular beam is identified as α and β, respectively, and is reflected in the out-of-plane stiffness matrix of the IATP element.

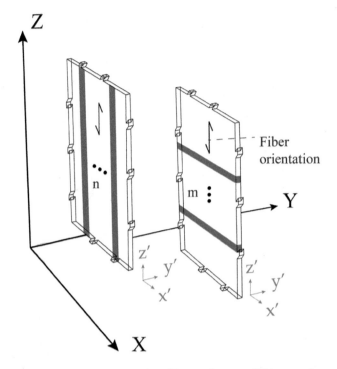

Figure 2.33 Division of plate into segmented beam elements ('n' beams along the fiber-parallel and 'm' beams along the fiber-perpendicular direction).

Figure credit: A. Rezaei Rad, et al. [18].

Assuming that the fiber-parallel and fiber-perpendicular directions have n' and m' strips, respectively, it follows that $\sum_{i=1}^{n'} \alpha_i + \sum_{j=1}^{m'} \beta_j = 1.0$ where $0 < \alpha_i, \beta_j < 1.0$. For complex assemblies with large numbers of DOFs, a numerical solution is required to compute α and β.

2.6.2.2 Forces and deformations, and virtual work principle

The out-of-plane behavior of each IATP is represented by that of the beam (strip) elements. To calculate the stiffness matrix of each beam (K_{strip}), the end nodes are indexed as [1] and [2] in Figure 2.34a, and the out-of-plane behavior is described using six DOFs in the GCS. The vector $D_{OP} = \{ u_1\ v_1\ \theta_1\ u_2\ v_2\ \theta_2 \}^T$ describes the kinematics associated with these DOFs when the element responds within the linear elastic range. The components of the vector D_{OP} are shown in Figure 2.34a. The terms u_1, v_1, u_2, and v_2 are the translations, and θ_1 and θ_2 are the rotations at the end nodes. The element force vector, F_{OP}, which corresponds to the vector D_{OP}, is then defined. Similar to the vector D_{OP}, the element force vector

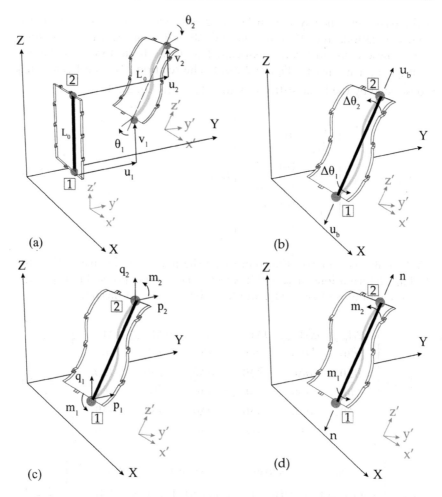

Figure 2.34 (a) out-of-plane displacement field, (b) out-of-plane force in the basic reference system, (c) undeformed and deformed configurations, and (d) free body diagram for out-of-plane bending about x' axis.

Figure credit: A. Rezaei Rad, et al. [18].

$F_{OP} = \{\, p_1 \; q_1 \; m_1 \; p_2 \; q_2 \; m_2 \,\}^T$ consists of six components, where p_1 and q_1 are the translation forces, and m_1 is the flexural moment associated with node [1]. Similarly, p_2 and q_2 are the translation forces, and m_2 is the flexural moment associated with node [2].

Given that the undeformed element has the initial length of L_0, the rigid body rotation of the element and the deformed length are denoted as ω and L'_0, respectively. The out-of-plane displacement field of a strip element about the local x' axis is described by the vector $u_{OP} = \{\, \Delta u_b \; \Delta \theta_1 \; \Delta \theta_2 \,\}^T$

at the basic reference system in Figure 2.34b, where $\Delta u_b = L'_0 - L_0$ is the axial elongation, and $\Delta\theta_1 = \theta_1 - \omega$ and $\Delta\theta_2 = \theta_2 - \omega$ are the nodal rotations. These components are associated with the basic forces of the element $f_{OP} = \{ n\ m_1\ m_2 \}^T$(Figure 2.34d). The relation between F_{OP} and u_{OP} is described by the basic stiffness matrix K_{basic}, in Eq. 2.24.

$$
K_{Strip} = B_{OP}{}^T.
\begin{bmatrix}
\dfrac{EA}{L_0} & 0 & 0 \\[2mm]
0 & \dfrac{4EI}{L_0} & \dfrac{2EI}{L_0} \\[2mm]
0 & \dfrac{2EI}{L_0} & \dfrac{4EI}{L_0}
\end{bmatrix}
.B_{OP}
\tag{2.24}
$$

Where B_{OP} is a matrix that represents the partial derivative of the out-of-plane deformations, u_{OP}, with respect to D_{OP} and expressed in Eq. 2.25, 'E' is the modulus of elasticity, and 'I' is the moment of inertia.

$$
B_{OP} =
\begin{bmatrix}
\dfrac{\partial\Delta l_{beam}}{\partial u_1} & \dfrac{\partial\Delta l_{beam}}{\partial v_1} & \dfrac{\partial\Delta l_{beam}}{\partial\theta_1} & \dfrac{\partial\Delta l_{beam}}{\partial u_2} & \dfrac{\partial\Delta l_{beam}}{\partial v_2} & \dfrac{\partial\Delta l_{beam}}{\partial\theta_2} \\[3mm]
\dfrac{\partial\Delta\theta_1}{\partial u_1} & \dfrac{\partial\Delta\theta_1}{\partial v_1} & \dfrac{\partial\Delta\theta_1}{\partial\theta_1} & \dfrac{\partial\Delta\theta_1}{\partial u_2} & \dfrac{\partial\Delta\theta_1}{\partial v_2} & \dfrac{\partial\Delta\theta_1}{\partial\theta_2} \\[3mm]
\dfrac{\partial\Delta\theta_2}{\partial u_1} & \dfrac{\partial\Delta\theta_2}{\partial v_1} & \dfrac{\partial\Delta\theta_2}{\partial\theta_1} & \dfrac{\partial\Delta\theta_2}{\partial u_2} & \dfrac{\partial\Delta\theta_2}{\partial v_2} & \dfrac{\partial\Delta\theta_2}{\partial\theta_2}
\end{bmatrix}
\tag{2.25a}
$$

$$
B_{OP} =
\begin{bmatrix}
-\cos(\omega) & -\sin(\omega) & 0 & \cos(\omega) & \sin(\omega) & 0 \\[2mm]
-\sin(\omega)/L'_0 & \cos(\omega)/L'_0 & 1 & \sin(\omega)/L'_0 & -\cos(\omega)/L'_0 & 0 \\[2mm]
-\sin(\omega)/L'_0 & \cos(\omega)/L'_0 & 0 & \sin(\omega)/L'_0 & -\cos(\omega)/L'_0 & 1
\end{bmatrix}
\tag{2.25b}
$$

Using Eq. 2.24, K_{strip} is then computed according to Eq. 2.26.

$$
K_{Strip} = \frac{\partial F_B}{\partial D_B} = \frac{\partial}{\partial D_B}\left(B_{OP}{}^T.f_{OP}\right) = B_{OP}{}^T.\frac{\partial f_{OP}}{\partial D_{OP}} + \frac{\partial B_{OP}{}^T}{\partial D_{OP}}.f_{OP}
\tag{2.26}
$$

Neglecting the higher effects $\left(\frac{\partial B_{OP}{}^T}{\partial D_{OP}}.f_{OP}\right)$, the stiffness matrix of a strip element can be expressed in Eq. 2.27.

$$
K_{Strip} = B_{OP}{}^T.\frac{\partial f_{OP}}{\partial D_{OP}} = B_{OP}{}^T.\frac{\partial f_{OP}}{\partial u_{OP}}.\frac{\partial u_{OP}}{\partial D_{OP}} = B_{OP}{}^T.k_{basic}.B_{OP}
\tag{2.27}
$$

Applying the virtual work principle, the external force vector F_{OP} is expressed by the internal force vector, f_{OP}, in Eq. 2.28.

$$W_{external} = \delta W_{internal} \rightarrow \delta(D_{OP}{}^T).F_{OP} = \delta(u_{OP}{}^T).f_{OP} = \delta(D_{OP}{}^T).B_{OP}{}^T.f_{OP} \quad (2.28)$$

2.6.2.3 Out-of-plane stiffness

After determining K_{strip} in Eq.2.26 and applying Eq. 2.27, the principle of superposition is used to obtain the stiffness matrix for the entire IATP element. This is taken as the sum of the stiffnesses of the IMAs and the strip elements along the fiber-parallel and fiber-perpendicular directions while accounting for the contribution of each strip element to the out-of-plane behavior, and it is expressed in Eq. 2.29.

$$K_{\text{out-of-plane, IATP}} = K_{\text{out-of-plane, Plate}} + K_{\text{out-of-plane,Connection}} \quad (2.29a)$$

$$K_{\text{out-of-plane, IATP}} = \left\{ \sum_{j=1}^{m} B_{OP,j}{}^T . \begin{bmatrix} K_{IMA,tns} & 0 & 0 \\ 0 & K_{IMA,flt} & 0 \\ 0 & 0 & K_{IMA,flx} \end{bmatrix} . B_{OP,j} \right\}$$

$$+ \left\{ \sum_{i=1}^{n} \alpha_i . K_{\text{Strip},i} + \sum_{j=1}^{m} \beta_j . K_{\text{Strip},j} \right\} \quad (2.29b)$$

Where $K_{\text{Strip},i}$ and $K_{\text{Strip},j}$ are the stiffness matrices for the beams parallel and perpendicular to the fiber orientation, respectively and α_i and β_i are the associated flexural stiffness contribution factors. Furthermore, $K_{IMA,tns}$, $K_{IMA,flt}$, and $K_{IMA,flx}$ capture the semirigid behavior of the wood-wood connections under tensile and flatwise shear loads and out-of-plane moments, respectively.

2.7 NUMERICAL SIMULATIONS AND DISCUSSIONS

After determining the kinematic DOFs and analytical behavior of timber plates with wood-wood connections, a numerical modeling approach is used to simulate the complex behavior of IATP structures. This is mainly due to the fact that the identification of load path, determination of the interaction of different IATP elements, quantifying the role of the boundary conditions and their influence on the global structural behavior, distribution of surface and point loads, and different scenarios for material orientation are the parameters that make the development of an analytical closed-form model for IATP structures impossible. Therefore, numerical models are generally preferred to be used for structural design and calculations.

Numerical models for the analysis of plate structures typically lie somewhere along a spectrum of complexity, as they are schematically shown

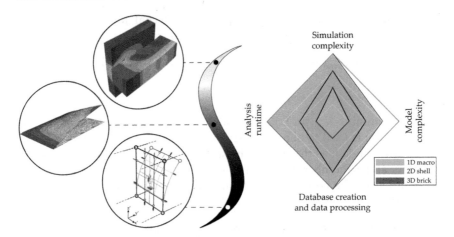

Figure 2.35 **Spectrum of complexity assumed for different numerical models exist in the design of IATP structures.**

Research credit: A. C. Nguyen, 2020 [4]; A. Rezaei Rad, 2020 [49]

in Figure 2.35. Finite element (FE) models with 3D solid (brick) elements, which can be viewed as existing on one end of that spectrum, are detailed, complex, and require a high level of sophistication. However, they provide the benefit of enabling an evaluation of the state of stress/strain across the plate thickness. In less complex models, the plate thickness is neglected, and 2D shell (surface) elements can be used to simulate the behavior of the plate. Both of the aforementioned approaches provide enhanced simulation capabilities to capture stress-strain response and local failure modes. In more simplified models, only beams and springs are utilized to simulate structural response. The primary goal of such models, which are generally referred to as macromodels, is to reduce the computational expense while maintaining the accuracy of the simulation.

2.7.1 Analysis and design of IATP structures using shell-based finite element (FE) models

Given that FE models with 3D solid (brick) elements are computationally expensive for large-scale structures with a large number of components simplified FE models using 2D shell (surface) elements are employed in the current study. This modeling approach can lead to considerable computational time savings, and it ensures that the behavior of timber plates is approximated with an acceptable degree of accuracy. Furthermore, shell elements can simulate thin elements with fewer mesh elements than that of the solid (brick) elements. Moreover, the use of solid (brick) elements in thin plate elements might lead to negative Jacobian errors, while this numerical error can be generally avoided by using 2D shell elements.

Connections in structures disturb the material continuity and are considered the weakest link in the load-carrying system. Therefore, it is essential to guarantee a reliable simulation of connections in numerical models. In FE models with 3D solid (brick), modeling the behavior of connections is generally a complicated procedure. The connection between the structural elements is established in these models by assuming a reliable constitutive behavior for the contact zones between two interconnected elements. Within this context, contact between the 3D solid elements is simulated by defining surface-to-surface interactions. Furthermore, mesh properties, damage evolution, failure modes, material characteristics, moisture content, fabrication effects, and imperfections and gaps are the most influential parameters that affect the behavior of the model, and they should be carefully defined and reflected in the numerical model database. On the other hand, modeling the behavior of connections in the FE models with 2D shell elements is more straightforward and conventional than that of the FE models with 3D solid (brick) elements. The use of shell elements and associated mesh properties will be discussed in Sections 2.7.1.1, 2.7.1.2, and 2.7.1.3.

Overall, the behavior of connections is generally simulated using springs elements in 2D shell-based FE models. These spring elements are then attached to the edges of the shell elements. In this case, the constitutive behavior of the spring elements, directly derived from the experimental tests, is assigned to the spring elements. Using simplified spring elements is relatively advantageous in simulating the semirigid behavior of connections, and it can realistically reflect the actual behavior of the component. Accordingly, rather than using fully rigid or fully pinned connectors, the design professional can ensure that the behavior of the structure is not over nor underestimated when the spring element is employed. Furthermore, this modeling procedure is fairly straightforward in structures with a large number of elements and connections, and they can be algorithmically (and parametrically) embedded in the development of the model database. The use of spring elements and other available alternatives will be discussed in Section 2.7.1.4.

The workflow corresponding to the design of timber plate structures with wood-wood connections using the shell-based FE model and spring elements is shown in Figure 2.36. This framework is designed by Nguyen [4] and Nguyen et al. [21] such that it covers a wide range of geometries ranging from standard shapes to freeform structures. The framework consists of different modules; nevertheless, it can be divided into two main clusters. In the first cluster, an automatic framework for generating the shell-based FE model corresponding to the geometry of a given structure is structured. To formulate this framework, an algorithmic CAD-to-FE data exchange module using a heterogeneous data exchange methodology [22] is introduced. As a result, the geometry of a structure is translated to the corresponding numerical CAE model for further FE analysis. The CAD-to-FE data exchange is also consistent with the computational tools developed for digital CAD modeling and fabrication planning. The output of the data exchange is then mapped

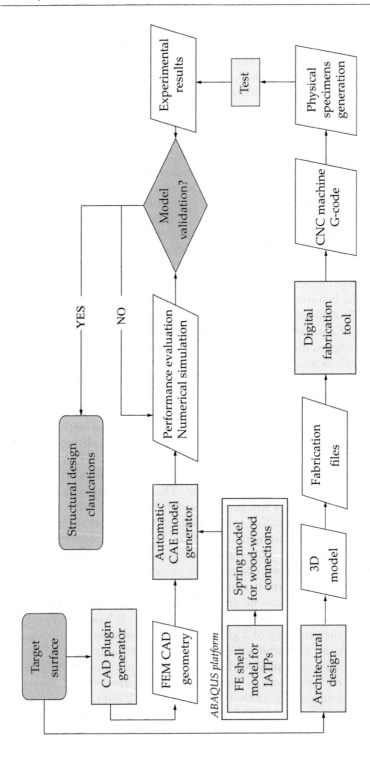

Figure 2.36 The framework developed for designing IATP structures consisting of the FE model generation from a design target surface and associated CAD-FE geometry translation, supported by small- and medium-scale experimental investigations.

to an FEA software using a custom scripting code. In the second cluster, the CAD-to-FE data exchange platform application is researched, and its scope of validity is assessed. Accordingly, the tool is validated against recent case studies by comparing the performance of the numerical model with the results obtained from the experimental tests. The primary aim of this step is to prove that an accurate structural design calculation is provided by the proposed framework. As another objective within the cluster, the structural performance and mechanical behavior of IATPs are detailed, and a comprehensive evaluation is provided for design professionals.

The modules included in the design of timber plate structures with wood-wood connections using the shell-based FE model are detailed in the following sections. The essential elements used to construct the FE model corresponding to an IATP structure are discussed. The automation of the FE model and CAD-to-FE data exchange scheme is then explained in detail. The FE simulation of the IATP structures is essentially carried out in the Abaqus computer-aided engineering environment. Accordingly, the modules are primarily written in Python language programming, being compatible with Abaqus computer-aided engineering software (Dassault Systèmes, Vélizy-Villacoublay, France). Two recent versions of Abaqus, i.e., 6.12 and 6.14, are used.

2.7.1.1 Modeling of orthotropic timber material

To define the shell elements corresponding to thin timber panels, the material properties of the LVL material should be defined first. To do so, the LVL material properties are simulated using orthotropic material properties. To simplify the modeling process and reduce the complexity of the simulation process, the laminated layers are summarized in a single layer, and the orthotropic properties are then assigned to this layer. The orthotropic material properties are described using Hooke's constitutive law shown in Eq. 2.30.

$$\varepsilon = C^{el}\sigma \rightarrow \begin{bmatrix} \varepsilon_{11} \\ \varepsilon_{22} \\ \varepsilon_{33} \\ 2\varepsilon_{12} \\ 2\varepsilon_{13} \\ 2\varepsilon_{23} \end{bmatrix} = \begin{bmatrix} \dfrac{1}{E_{11}} & \dfrac{-v_{12}}{E_{22}} & \dfrac{-v_{31}}{E_{33}} & 0 & 0 & 0 \\[2mm] \dfrac{-v_{12}}{E_{11}} & \dfrac{1}{E_{22}} & \dfrac{-v_{32}}{E_{33}} & 0 & 0 & 0 \\[2mm] \dfrac{-v_{13}}{E_{11}} & \dfrac{-v_{23}}{E_{22}} & \dfrac{1}{E_{33}} & 0 & 0 & 0 \\[2mm] 0 & 0 & 0 & \dfrac{1}{G_{12}} & 0 & 0 \\[2mm] 0 & 0 & 0 & 0 & \dfrac{1}{G_{13}} & 0 \\[2mm] 0 & 0 & 0 & 0 & 0 & \dfrac{1}{G_{23}} \end{bmatrix} \begin{bmatrix} \sigma_{11} \\ \sigma_{22} \\ \sigma_{33} \\ \sigma_{12} \\ \sigma_{13} \\ \sigma_{23} \end{bmatrix} \quad (2.30)$$

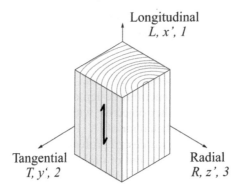

Figure 2.37 **Three principal directions used to define the orthotropic material proper-**
ties of timber LVL boards.

Where ε is the material strain, σ is the effective stress associated with the nondamage phase of the material, and C^{el} is the elastic compliance matrix.

Timber material consists of three main directions according to its fiber. The first direction is parallel to the fiber orientation. The second direction is perpendicular to the fiber orientation and tangential to the timber rings. The third direction is also perpendicular to the fiber orientation and radial to the rings. In the current study, the architectural design and fabrication process of IATP elements are such that the fiber orientation lies along the longitudinal dimension of the timber plates, as is shown in Figure 2.37.

2.7.1.2 Timber plates

The design of the IATP structures is centered around this assumption that the timber plates are planar, having no out-of-plane curvature. Furthermore, since a given IATP structure consists of planar segmented timber plates, each timber plate is simulated with an individual FE shell-based model. With this methodology, the geometrical complexity, orthotropic material properties associated with each timber plate, boundary conditions, and different load cases are readily defined and implemented in the engineering analysis.

Within the context of the shell-based FE approach, the numerical model corresponding to timber plates is defined using the principles of the plate and shell theory. For the sake of shell element definition, the actual geometry of a timber plate is simplified such that the plate thickness is not represented in the numerical model since it is significantly smaller than the two other dimensions. Accordingly, the midsurface of each timber plate is defined as the representative of the plate in the numerical model. Midsurface elements can be either simulated in stand-alone FE simulation software or generated within a CAD environment and transformed to the FE platform. Given that typical IATP structures potentially consist of hundreds of timber plates, an

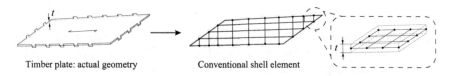

Timber plate: actual geometry   Conventional shell element

Figure 2.38 **The use of midsurface to represent timber plate and build the FE model with conventional shell element.**

Figure note: The geometry is regenerated based on the existing data and the framework investigated and developed by Nguyen [4] and Nguyen et al. [21]

algorithm is developed to automatically process the midsurface of timber plates and map the data to the FE platform. This process will be extensively discussed in Section 2.7.1.

Using the midsurface associated with each timber plate, a conventional shell element is used to simulate the behavior of timber panels (Figure 2.38). In the simulation methodology, the reference surface is discretized using the dimensions of the plate and its normal vector. Furthermore, because timber plates are essentially planar, the initial curvature required to construct the conventional shell element is set to be zero. Given that the dimension of each timber plate is reduced from 3D brick to 2D surface, the thickness of each shell element is analytically reflected in the cross-sectional properties of each shell element. Therefore, it is assumed that the interface of inner laminated layers is rigid, and the laminated layers behave like a monolith object. Accordingly, the interlaminar kinematic is not covered in the numerical shell model. The geometric order of the mesh elements is represented by quadratic (second-order) element types. Using conventional shell elements, three translational and two in-plane rotational degrees of freedom associated with each finite element of the shell model are taken into account. However, because the plate thickness is reflected as a mathematical value instead of modeling the thickness, the prediction of transverse shear strain and associated stress and the state of the force-deformation across the plate thickness are not taken into account in this modeling strategy. Besides, since second-order reduced-integration conventional shell elements are used, hourglass control is not required during the design process.

Within the context of digital fabrication constraints, CNC milling tools cause notch holes around the tenons of a timber plate. Using the shell elements, the notches are neglected during the simulation process, and they are implicitly reflected in the force-deformation or moment-rotation behavior of the wood-wood connections instead, which will be described in Section 2.7.1.4. Accordingly, the simulation complexity is considerably reduced since there is no need to provide refined mesh for the notch regions that potentially have sliver faces, short edges with very high curvature.

2.7.1.3 Mesh properties and modeling uncertainty

After defining the geometry of each part, it is necessary to discretize it to solve the system of equations using the FE method. The accuracy of the FE analysis heavily depends on the mesh configuration, the associated seed size, the mesh type, and the corresponding geometrical properties used for the shell elements. It is within this context that a relevant mesh size should be adopted in the simulation process. A refined mesh can incrementally increase the computational expense, however, increasing the mesh size does not necessarily lead to accurate results. When using a coarse mesh size for elements, a zeroenergy mode of response is prone to occur in the numerical simulation, which causes zero stiffness for the mesh element. Given this consideration, the mesh properties are one of the main sources of modeling uncertainty in designing IATP structures.

A numerical simulation is carried out to assess the sensitivity of the response of timber plates to the mesh size. The numerical analysis was performed on a single quadrilateral timber shell element. The element was subjected to a point load with the intensity of F = 6.47 kN at one single wood-wood connection. This level of load corresponds to the ultimate limit state of a single wood-wood connection given in Table 2.1. This ensures that the mesh sensitivity analysis is valid within the linear elastic range of responses and can be employed in the design process with an acceptable degree of reliability. The connections at the other edge side of the shell element are fully restrained. The overall geometry of the IATP element is shown in Figure 2.39a.

The tensile translational force associated with the wood-wood connection, to which the point load is applied, is considered as the intensity measure and called F_{Tensile}. The computation of F_{Tensile} is described in Eq. 2.31.

$$F_{tensile} = \sum_{j=1}^{m} SF_1 \delta_j \tag{2.31}$$

Where SF_1 is the sectional force component in the wood-wood connection per unit width and along the tensile direction, and δ_j is the length of the mesh used in the jth step.

The sectional force component is expressed in terms of the axial stress values that appeared along the section thickness. As such, the total amount of the applied force (F) is expressed in terms of the axial stress values, mesh size, and lever arm. This is expressed in Eq. 2.32.

$$F_{tensile} = \sum_{j=1}^{m} \left(\int_{-z}^{-z} \sigma_{11,j}(z) dz \delta_j \right) \tag{2.32}$$

Where $\sigma_{11,j}$ is the axial stress value of the jth step, and it is integrated along the section thickness, z.

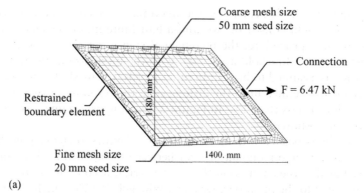

Coarse mesh size
50 mm seed size

Connection

F = 6.47 kN

Restrained
boundary element

Fine mesh size
20 mm seed size

1180. mm

1400. mm

(a)

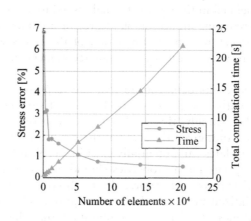

(b)

Figure 2.39 Mesh sensitivity analysis (a) geometry, boundary condition, and applied loads, and (b) mesh convergence, including normal stress, axial force, and total computational time.

Figure note: The geometry in (a) is regenerated based on the existing data and the framework investigated and developed by Nguyen [4] and Nguyen et al. [21] and Rezaei Rad, 2020 [49].

Computing the amount of F_{Tensile} for different mesh sizes, the error for each case is defined as the difference between $F_{tensile}$ and the applied load (F), and it is shown in Eq. 2.33.

$$\text{Error}(\%) = \frac{F - F_{tensile}}{F} * 100 \tag{2.33}$$

The mesh size domain ranges from 2.5 mm to 29.0 mm in the sensitivity analysis. The minimum and maximum extremes for the mesh size were chosen to be compatible with the length of the connection, which was 72.5 mm. In this case, using the mesh size of 2.5 mm, the edge of the connection consists of 29 finite mesh elements. For the maximum extreme, the edge of the connection would consist of two-and-a-half finite mesh elements. Within the refinement procedure, the incremental step is set to be 2.5 mm. The acceptance error threshold associated with the mesh size in the numerical simulation is assumed to be 1%. In other words, the mesh density within a shell element is refined until the difference in the response of two consecutive solutions is less than 1%. In Figure 2.39a, a hybrid refine-coarse mesh distribution is shown.

The results associated with the mesh size of 2.5 mm, shown in Figure 2.39b, indicate the minimum amount of error between the applied load (F) and the observed load in the connection ($F_{tensile}$). The error, according to Eq. 32, and corresponding to the mesh size, was approximately 0.52%. Furthermore, the coarse mesh size defined in the sensitivity analysis led to an error of approximately 6.8%, representing an acceptable degree of accuracy. However, for large-scale IATP structures, the accumulation of the errors would lead to underestimated forces. Accordingly, it is not recommended to use such a mesh size.

The FE analysis results demonstrate that the response of timber shell plates discretized with a seed size of 20.0 mm and smaller tend toward a unique value. Accordingly, this seed size is considered a refined mesh size. On the other hand, given the scale of the structure and the fact that hundreds of timber plates can potentially exist in a typical IATP structure, relevant and appropriate mesh properties should be adopted to minimize the computational expense. Therefore, the mesh convergence study suggests that an optimized mesh simulation is required. The aim of adopting the optimized mesh modeling is to guarantee that the results of the numerical model are adequate as well as reduce computational resource use. Given these two constraints, the refined mesh size (20 mm) is applied only to those areas in the timber shell element that is close to the wood-wood connections. This is mainly due to the stress and strain concentration being considerably dense in such areas, and a refined mesh is required to ensure that the response is accurate. For areas that are not close to the connection regions, a relatively coarse mesh with a size of 50.0 mm is considered. Figure 2.39a demonstrates the refine and coarse mesh configuration distributed on a shell element. Employing the hybrid refined coarse mesh configuration reduces the computational expense by half while reducing the accuracy only by 0.4%. The time required by the machine is also shown in Figure 2.39b. It is worth noting that the machine used to carry out the sensitivity analysis had Intel® Core™ i7-4800MQ CPU @ 2.7GHz with 16 GB of RAM.

2.7.1.4 Simulation of through-tenon wood-wood connections

The numerical simulation of wood-wood connections is the most important part of the workflow shown in Figure 2.36. This is mainly due to the fact that the midsurface shell elements do not have any shared edges or overlaying nodes, and they are connected only by means of wood-wood connections. Understanding the role of the connections and their contribution to the global performance of the system helps the design professional to realistically simulate them in the computational platform. Given the structural system of IATP structures described in Section 2.2 and Figure 2.5, two types of connections exist in such structures; (a) dovetail and (b) through-tenon wood-wood connections. Each of these categories has a different representation and mechanical behavior. While dovetail joints are primarily used to connect cross plates, the through-tenon joints are used to join the top and bottom plates to the cross plates. The importance of each connection type and the simulation technique used to define the corresponding numerical model is elaborated in the following paragraphs.

In this paragraph, the focus is on the dovetail connections in IATP structures and how they will be simulated in the FE models. According to the geometrical position of the cross plates in IATP structures (Figure 2.5), the dovetail connections are mainly subjected to in-plane flexural behavior (Figure 2.28). Furthermore, dovetail connections provide a continuous connection by joining the shared edge of two inserting cross plates. This feature entirely interlocks the cross plates and provides a rigid tie between them. Hence, the dovetail connections are considered rigid. This assumption is supported by the experimental performance of the IATP structures discussed in Chapter 3, where these dovetail connections remained almost intact and maintained the consistency of the cross plates within each box. In terms of the FE simulation process, a rigid tie constraint represents the dovetail connections and bonds the shared edge of the cross plates (CL_i and CT_i in Figure 2.5) within each box together. Therefore, the midsurfaces corresponding to the plates CL_i and CT_i are imported to the FE platform as one integrated and individual part (Figure 2.40). Consequently, an easy

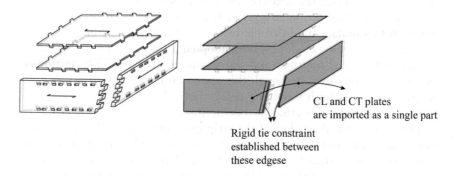

CL and CT plates
are imported as a single part

Rigid tie constraint
established between
these edgese

Figure 2.40 Rigid tie constraint established between CL_i and CT_i.

Figure note: The geometry is regenerated based on the existing data and the framework investigated and developed by Nguyen [4], Nguyen et al. [21], and Rezaei Rad, 2020 [49]

mesh transitioning between the cross plates per each-sided box is offered, and the two cross plates remain tied throughout the analysis.

The simulation process of through-tenon connections is completely different from the dovetail connections, and it is extensively discussed in this and the next following paragraphs. Before explaining the simulation technique used for the through-tenon connections, the necessity of the use of such joints is clarified. This would help to distinguish the role of through-tenon connections with the dovetail connections. In detail, the architectural design of IATP structures requires that the dihedral angle between two adjacent top plates is considerably large. This is mainly due to the fact that such structures are freeform, and they occupy a large distributed surface area in space. Therefore, and to follow the target surface and satisfy the set curvature, the dihedral angle should be large. For instance, the dihedral angle between the plates T_1 and T_2 is shown in Figure 2.42a. For most of the adjacent top plates in free-form structures, this angle converges to $180°$. Accordingly, a side-to-side contact exists between the adjacent top layers. Given that the side-to-side contact cannot establish a connection between the two adjacent plates, an interface timber plate is required to join the elements. Within this context, the cross plates (CL_i and CT_i in Figure 2.5) are introduced to the design process. These cross plates are meant to connect the top plates of a given IATP structure. Given these constraints, Through-Tenon joints are accordingly used to establish the connection. Based on the assembly technique and the role of the through-tenon joints to eliminate the effect of the large dihedral angle, the through-tenon joints govern the local and global behavior of IATP structures. This was also observed in the performance of the physical experiments discussed in Chapter 3.

As mentioned at the beginning of Section 2.7.1, spring elements are generally used in shell-based FE simulations to simulate the behavior of connections. Nevertheless, there is another numerical approach in the simulation of connections. In this approach, a strip of fictitious continuous material is used along the edges of connected shell elements to simulate the stiffness of the connection. Bagger [23] initially suggested this modeling approach to simulate the isotropic behavior of connections in glass plate shell structures. Stitic et al. [24] later adopted a similar approach to simulate the behavior of multitab-and-slot wood-wood connections in timber folded plate structures. Although this approach has been verified against experimental tests with satisfying results [24], the use of spring elements is more computationally efficient. Furthermore, the use of spring elements is more convenient than that of the strip elements, especially in complex geometries. Moreover, spring elements can be readily implemented in conventional numerical FE packages, and in that sense, the use of spring elements is more straightforward than that of the strip elements. The simulation of the wood-wood connections using the spring and strip elements is shown in Figure 2.41. Furthermore, the use of spring elements to simulate the through-tenon wood-wood connections is elaborated in Figure 2.42b.

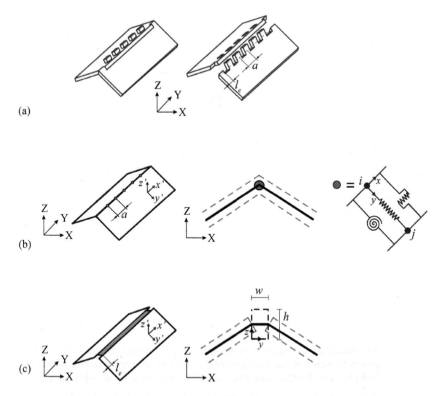

Figure 2.41 Simulation of connections in timber plates, (a) multitab-and-slot wood-wood connection with through-tenon geometry, (b) the use of spring elements to simulate each through-tenon connection, and (c) the use of strip element to simulate the entire through-tenon connections.

Figure note: The geometry is regenerated based on the existing data and the framework investigated and developed by Stitic et al. [24], Nguyen [4], Nguyen et al. [21], and Rezaei Rad, 2020 [49]

The kinematic degrees of freedom associated with the 3D continuum geometry of the joint region is idealized using a two-node link element in this study. The element consists of six different springs that capture the tensile, edgewise (in-plane), and flatwise (out-of-plane) force-deformation and the flexural and torsional moment-rotation behavior of the joint. Figure 2.42c shows the joint region and the equivalent spring elements. Given that each node has six DOFs, the two-node link element has 12 DOFs.

Force-deformation or moment-rotation behavior resulting from experimental tests is used to compute the translational and flexural stiffnesses of the spring elements. Recalling Figure 2.42c, two-node link elements, each with six subspring elements, are used to connect timber shell elements in the FE numerical model. Each subspring element has its mechanical stiffness; therefore, the subsprings are uncoupled. In other words, it is assumed

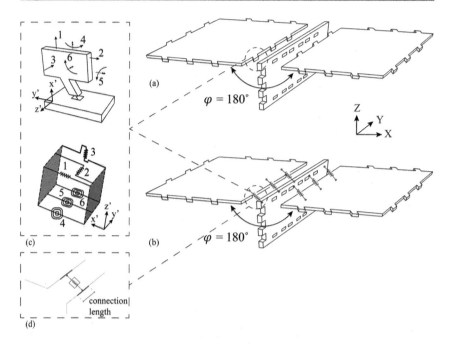

Figure 2.42 The dihedral angle between the neighboring top plates, the use of spring elements to simulate the through-tenon wood-wood connections, and identified DOFs with equivalent two-node link element with six subsprings.

Figure note: The geometry is regenerated based on the existing data and the framework investigated and developed by Nguyen [4], Nguyen et al. [21], and Rezaei Rad, 2020 [49]

that the behavior of the connection in one direction is completely independent of the response in the other five directions. Given this, the general form of the element stiffness matrix associated with each two-node link element is expressed in Eq. 2.34.

$$K_{IMA} = \begin{bmatrix} k_{tns} & 0 & 0 & 0 & 0 & 0 \\ 0 & k_{edg} & 0 & 0 & 0 & 0 \\ 0 & 0 & k_{flt} & 0 & 0 & 0 \\ 0 & 0 & 0 & k_{tor} & 0 & 0 \\ 0 & 0 & 0 & 0 & k_{flx-y'} & 0 \\ 0 & 0 & 0 & 0 & 0 & k_{flx-x'} \end{bmatrix} \qquad (2.34)$$

Where K_{IMA} is the stiffness matrix with six uncoupled components for element representation, k_{tns} is the tensile stiffness denoted as DOF#1 in Figure 2.42c and described in Figure 2.11 and Figure 2.12, k_{edg} is the edgewise (in-plane) stiffness denoted as DOF#2 in Figure 2.42c and described in Figure 2.13, k_{flt} is the flatwise (out-of-plane) stiffness defined as DOF#3 in Figure 2.42c and described in Figure 2.14, k_{tor} is the torsional stiffness of the joint corresponding to DOF #4 in Figure 2.42c, $k_{flx-y'}$ is the flexural stiffness of the connection about the local axis y' denoted as DOF #5 in Figure 2.42c and described in Figure 2.15, and $k_{flx-x'}$ is the in-plane flexural stiffness about the local axis x' denoted as DOF #6 in Figure 2.42c. Preliminary numerical simulations and experimental tests demonstrated that k_{tor} and $k_{flx-x'}$ are considerably rigid. Accordingly, a rigid value is assigned to the torsional behavior, k_{tor}, and in-plane flexural stiffness $k_{flx-x'}$.

The use of spring elements to reflect the joint region can also take into account the effect of notches in the wood-wood connections. In fact, the connections and plates are one unit in IATP structures. In other words, because no additional connectors such as screws or fasteners are employed, the joints and the associated plate cannot be separated. It was within this context that the physical experiments included both the joints and the neighboring region of the plate in the test specimens. With this type of experiment, the behavior of what is referred to as the 'joint region' was examined, where the notch holes were included in the test process. Consequently, the effect of notches is explicitly included in the initial stiffness from an idealized backbone curve (Figure 2.27) and thus, reflected in the spring elements. As a result, the use of spring elements to explicitly describe the behavior of the joint region can account for the local behavior of IATPs.

In terms of the FE simulation, connector elements are used to represent the springs and model discrete and point-to-point physical connection between the shell elements. Different approaches exist to implement the connector elements and provide joinery for timber shell elements in a FE model. One approach is to use mesh-independent point fasteners. In this approach, the point-to-point connection between surfaces is provided. While this approach is very efficient in connecting two or multiple surfaces, connecting two edges of timber shells seems problematic with numerical instabilities with this approach. Furthermore, given the geometry of IATPs where several connections potentially exist in the edge of a timber shell, over constraint errors are prone to occur while using this modeling technique. In other words, multiple inconsistent (conflicting) kinematic constraints might be applied to a single degree of freedom without the user's notice, which leads to numerical errors. Another approach is to adopt multiple spring elements distributed along the length of the wood-wood connection to provide the connection. This method is generally used to avoid stress/strain concertation and enable the shell element to distribute the forces coming from the spring elements uniformly. Although this approach reduces numerical instability caused by a dense stress/strain concertation, and the numerical model is less ill-posed than the mesh-independent point fasteners, it requires a large number of spring elements in a FE model, and thus, it augments computational cost.

One of the most efficient techniques to simulate connections in shell structures without considerable stress concentration and numerical instability is to use only one spring element with surface-based coupling constraints. In the current investigation, this modeling approach is used to simulate the through-tenon wood-wood connections. In this approach, a single spring element is used to connect a single node of one shell element to a single node of the neighboring shell element. These two nodes are located in the middle of the connection, and they are recognized as the reference nodes. By employing the coupling interaction, a constraint between the reference node and the region on the edge of the shell over which the connection transfers loading is provided. Therefore, a rigid constraint between the master node and the coupling nodes located on the edge of the shell element is provided. The modeling approach is schematically shown in Figure 2.42d. In this approach, the load applied by the spring element is uniformly distributed over the length of the connection. While the uniform load distribution is the most straightforward and efficient approach, other weighting methods enable the load to be monotonically decreased with radial distance from the reference node.

Per each connector, a local coordinate system is defined, enabling the definition of the mechanical properties. The local x' axis is determined from the nodal geometry, and it is parallel to the vector connecting the nodes. The local y' axis always represents the in-plane behavior of the connection, and the local z' is defined by the cross product between the local x' and y' vectors. The end nodes of each spring element are attached to a node of the shell element. Figure 2.42d shows the spring element with the surface-based coupling constraint method. K_{IMA} defined in Eq. 2.34 is used to define the mechanical properties of the spring element.

Identification of the location of each wood-wood connections, defining the corresponding local coordinate system, assigning the relevant mechanical properties, coupling the nodes that belong to each wood-wood connection, and connecting the identified connections to the corresponding midsurface shell elements are carried out through the development of an automatic algorithm within the CAD environment. The workflow associated with the design framework is extensively explained in Section 2.7.1.

2.7.2 CAD-to-FE data exchange and transformation

Since architectural models and the corresponding numerical models are constructed in different environments, the needs to develop methodologies to convert CAD data into FE models is inevitable. A typical CAD-to-FE exchange integrates multiple types of parameters, including the geometric features of the real physical system and how these features will be represented in the CAD and FE models, the scale of the CAD model, and the complexity and level of detail represented in the CAE model that is appropriate for the problem at hand.

Overall, CAD-centric and CAE-centric design philosophies are used to develop the exchange algorithm and the associated system architecture. In CAD-centric design, an iterative procedure is used to develop, improve, and refine the design object. Once the CAD model is developed, it is then idealized and converted into the formats needed for the structural analysis. The idealization process includes well-known transformations such as detail removal [25], dimensional reduction [26], and nonmanifold topological modeling [27]. On the other hand, in CAE-centric design, engineering analysis is incorporated in the early stages of the design process [28]. The CAE-centric design philosophy is used to analyze the idealized design object and improve the performance prior to CAD development. The associated CAD models are established after the engineering design by adding details to the abstracted CAE model. In the CAE-centric design approach, the design object is essentially developed using midsurface elements, and therefore, it is the ideal design strategy. The CAE-centric design approach enables both the architectural designer and simulation engineer to feed the project details and accomplish the design model gradually. However, in almost all design cases corresponding to IATP structures, the CAD-centric design scenario is adopted. This is mainly due to the dominant role of the architectural design, digital fabrication limitations, sequential assembly constraints, and interlocking and joinery of timber plates. Therefore, the primary aim is to use the CAD-centric approach and offer an integrated framework to systematically create a unique geometric model for both CAD and CAE analyses. The features of both models are stored in a 'master model.' The system architecture of the master model combines elements of feature-based solid models and nonmanifold topological models. Accordingly, the data structure associated with the CAD and CAE formats is simultaneously exchanged and updated.

The preparation of a CAE model requires that the original CAD model is considerably simplified. The simplification is primarily centered around dimensional reduction. This technique, which is used to convert 3D solid CAD elements to 2D midsurface shell elements and wood-wood connections to two-node link elements with springs, was extensively explained and discussed in Section 2.7.2. In addition, other simplifications are applied to the original CAD model. In particular, notch holes resulting from digital fabrication are neglected, boundary conditions are modeled with simplified pinned or fixed constraints, and embedment pressure between neighboring plates are not included in the CAE model. Moreover, to implement the connections between the timber plates, the topology of each timber plate is imprinted and its actual dimension is modified. In this regard, the midsurface of each timber plate is extended by $\frac{t}{2} - 0.5$ mm, where t is the plate thickness, in order to provide a microgap between the adjacent plates. Given the different workflows required to construct a CAE model, it appears that the geometry cleanup procedure and defeaturing the original CAD model is labor intensive manual work for IATP structures. Therefore, a general-purpose preprocessor is introduced to separate the unwanted features and also to read the real CAD geometry and map the data required for the FE analysis to the CAE platform.

The most challenging roadblock to automate the CAE model using the FE method is to adopt an appropriate data transformation scheme to ensure an adequate representation of the actual structure. There are two main methodologies to adopt CAD-to-FE data transformation. In the first method, the simplified geometry of a given object is generated in the FE simulation software platform using its default CAD-integrated tool. This method is quite efficient for simple geometry. On the other hand, to set up the FE model corresponding to a complex geometry with a large number of elements, a parametric data transformation algorithm is generally developed inside the native CAD program. Since IATP structures fall into this category, a modular and systematic scheme is developed to transform the geometry from the native CAD platform to the target CAE FE platform. Although this process depends highly on the target CAE platform, the algorithm developed herein is formulated to be compatible with different computational environments.

The generation of the CAE FE model requires that the structural components within a custom-defined structure are classified. Accordingly, two main classes, called *box* and *plate*, are defined as the basis of the data structure. Each plate and box component within the data structure is labeled with an exclusive and identical index. Furthermore, the connectivity between the adjacent plate and boxes and the base planes for plate elements are stored in a hierarchical way. To build the CAE model using the CAD model, the same discretization method used for the geometry generation is used. The geometry of each timber plate in the FE model is represented by four points located at the corner of the plate element and connected with line elements. This, in particular, builds a polyline that represents the geometry of the timber plate. Following the meshing strategy described in Section 2.7.1.3, each perimeter polyline is divided into two parts: an outer part where a file mesh is applied and an inner part where a coarse mesh is applied. Furthermore, the algorithm determines the location of the joints and model them with a series of rectangular elements perpendicular to each edge.

To compute the contour geometry for timber plates and generate the CAD-FE geometry, plane-plane and plane-line intersection methods are used. Using these intersection methods shown in Figure 2.43a, each plane – called P_i – within a box is computed using the base mesh edge direction and the normal vector available in the CAD model. Next, the contours associated with the top and bottom plates are generated. This is mainly done by identifying $Line_{i,i+1}$ located at the intersection of the plane P_i and P_{i+1}. $Line_{i,i+1}$ is then intersected with top and bottom planes, denoted as P_T and P_B. The process is illustrated in Figure 2.43b. Similar to the top/bottom contours, the contours associated with the cross longitudinal and cross transverse plates are computed. In this step, the lines L_{0T}, L_{2T}, L_{0B}, and L_{2B} are computed at

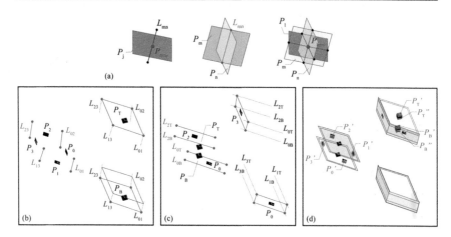

Figure 2.43 (a) Plane-plane, plane-plane-plane, and line-plane intersection methods used in the CAD-FE data exchange algorithm. Development of the contour lines associated with the (b) top and bottom plates, (c) cross longitudinal and cross transverse plates, and (d) line offsets used to define the coarse and fine mesh.

Figure note: The geometry in (a) is re-generated based on the existing data and the framework investigated and developed by Nguyen [4] and Nguyen et al. [21]. The geometry in (b) to (c) is reprinted from A. C. Nguyen, P. Vestartas, and Y. Weinand, Design framework for the structural analysis of free-form timber plate structures using wood-wood connections, in: Automation in Construction, vol. 107, Elsevier, 2020.

the intersection of the planes P_0, P_2, P_T, and P_B (Figure 2.43c). Consequently, these lines are connected, where they make a polyline that represents the cross plates. After determining the polylines, they are converted to surface elements. According to the mechanical/structural assumptions, each surface element should be divided into two main areas: an inner region associated with a coarse mesh and an outer region associated with fine mesh. The outlines of this region are determined by offsetting the neighboring planes along the normal direction of the plate and making an intersection between these planes and the baseline (Figure 2.43d).

After the identification of surface elements, the location of the through-tenon joints is determined. In this regard, interpolated rectangular-shaped areas are identified by the algorithm to represent the through-tenon wood-wood connections. These rectangular-shaped areas are essentially perpendicular to the edges of the polylines shown in Figure 2.43b-c. These areas are located in the outer region of each surface element, and they are shown in Figure 2.44a for a given timber box. Generally, the orientation of timber fibers is parallel to that of the length of each polygon (vector e_1 in Figure 2.44b). The local coordinate system for each joint is shown in Figure 2.44b. Given the ith timber plate, the direction of the plate edge and the vector normal to the plate are defined as e_1 and n_i, respectively. Accordingly, the orientation of each joint is determined by

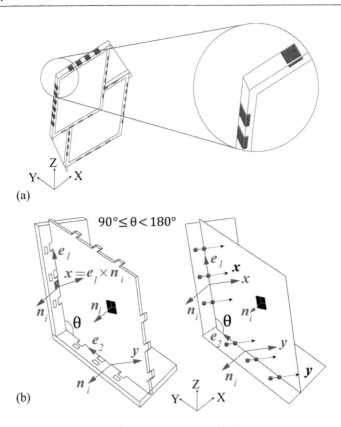

Figure 2.44 (a) Identifying the through-tenon wood-wood connections, and (b) local coordinate system for timber joints.

Figure note: The geometry in (a) is re-generated based on the existing data and the framework investigated and developed by Nguyen [4], Nguyen et al. [21]. The geometry in (b) is re-generated based on the existing data and the framework investigated and developed by A. Rezaei Rad, et al. [29]

computing the cross product of the vectors e_1 and n_i ($e_1 \times n_i$). After determining the three principal orientations, rectangular areas are generated using the plane-line intersection.

The CAD-to-FE data exchange algorithm is developed in a way that does not affect the CAD geometry robustness. In other words, the algorithm guarantees that the simplifications done to prepare the CAE model does not alter the original CAD model. Nevertheless, any changes in the CAD model synchronously change and update the CAE model. Although the data exchange principle can be applied to various computational platforms, the Rhinoceros .NET Software Development Kit (SDK), called RhinoCommon [30] and Rhino.Python [31], is employed within the Rhinoceros® version 6.0 environment to construct the plugins and components for the CAD-to-FE data transformation and FE geometry manipulation in this study. Given that Python is compatible with the CAE Abaqus environment as well, the geometry required for the FE analysis can be directly exported from Rhinoceros and mapped to Abaqus. The geometry processed and prepared for the FE analysis is exported from the CAD environment and introduced to Abaqus using its application programming interface (API), called Abaqus Scripting Interface. The interface allows full control over the model database, such as parts, materials, loads, steps, analysis handlers, and also output database.

Using the CAD-to-FE data exchange, the FE model generation initiates with defining a bounding domain, including the entire elements of the structure according to the boundaries of its global geometry. This facilitates the simulation process by parametrically assign the mechanical properties. For instance, by defining a boundary domain that includes the entire geometry within itself, the material and cross-sectional properties can be readily assigned to the entire elements at once. Next, the processed geometry, required by the Abaqus FE solver, is exported from the Rhinoceros 3D platform using an ACIS-format file (.SAT). Each component associated with the datastructure is mapped to the FE Abaqus software along with its position in 3D space. The top and bottom plates are imported as individual parts. However, reiterating that the dovetail connections are assumed to behave rigidly, the cross longitudinal and cross transverse plates are imported as one monolith object.

Given the geometry that is compatible with the FE solver, material properties associated with LVL timber material are defined and assigned to each individual surface element. The orthotropic material properties used in this step are provided by the manufacturer catalog. For instance, for Beech BauBuche LVL material, the mechanical properties of the timber panel are summarized in Table 2.2. Each timber plate has its fiber orientation in 3D space. Therefore, the discrete material orientation method is used to define a spatially varying orientation for each timber plate. In this technique, the centroid of the topology of each part is picked up, and a right-handed Cartesian coordinate system, including the information about the fiber directions, is accordingly defined. The primary axis is always parallel to the longitudinal direction of the plate, which is also parallel to the fiber direction. Furthermore, the normal axis is aligned with the local z' direction, which is perpendicular to the surface face, as shown in Figure 2.45a.

Table 2.2 Material properties of beech BauBuche [12-13].

Symbol	Description	Value (N/mm²)
E_0	Modulus of elasticity, fiber-parallel	13,200
E_{90}	Modulus of elasticity, fiber-perpendicular	2200
G_0	Shear modulus, fiber-parallel	820
G_{90}	Shear modulus, fiber-perpendicular	430

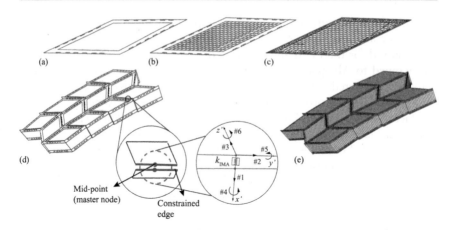

Figure 2.45 CAD-to-FE data exchange platform (a) importing the processed geometry compatible with the CAE Abaqus platform, (b) domain subdivision, generation of coarse mesh, and identification of the connection positions, (c) assignment of timber material, the definition of section properties for shell elements with an associated local coordinate system, generation of the refined mesh, and partitioning of the edges and associated mesh generation, (d) assembly of the instances and parts, and generating the wood-wood connections, (e) CAE model prepared for structural analysis.

Figure note: The geometry is regenerated based on the existing data, and the framework investigated and developed by Nguyen [4], Nguyen et al. [21], and Rezaei Rad, 2020 [49].

To define shell elements for each imported part, the material properties define a homogeneous section property. The thickness, thickness integration rule, and the number of integration points through the thickness are also specified in this step. General-purpose thin shell elements are used to account for finite membrane strains, as well as large rotations. In particular, the finite-strain shell element with four nodes, S4R, is considered as the primary element. This type of shell element has six DOF, and it enables thickness changes, which will lead to a realistic evaluation of the performance of the structure. The S4R shell element uses lower-order integration, which is also referred to as uniformly reduced integration, to compute the element stiffness. In this technique, only the geometrical center of the shell element is included in the integration process. One of the main benefits of using S4R shell elements is that numerical instabilities such as transverse shear and membrane locking and unconstrained hourglass modes are less prone to occur.

Partitioning each surface element into two main parts is carried out in this step. In detail, each surface element is divided into interior and exterior regions shown in Figure 2.45a. The interior region is subjected to a coarse mesh, while the exterior region is subjected to a fine mesh. The seed size adaptation corresponding to the coarse and fine mesh sizes is discussed in Figure 2.39. Quad-dominated element is adopted as the mesh shape. Furthermore, since the topology of each surface element is different from one to another, a free meshing technique is used. In this technique, a flexible meshing approach is offered mainly because pre-established mesh patterns are not employed. Given that wood-wood connections are located around the edges of plate elements, the advancing front algorithm is used to generate quadrilateral mesh elements at the edges of plates and then continue spreading the mesh systematically to the interior region. The advancing front algorithm is used for both coarse and fine meshes. Since timber surface elements do not have any holes, sliver faces, or tiny edges, the advancing front algorithm can distribute mesh elements with an acceptable degree of accuracy. A timber plate with coarse interior mesh is shown in Figure 2.45b. Next, the regions corresponding to the Through-Tenon wood-wood connections are identified. Each region is represented by a small portion of the area of the surface element. The length of the region lies along the perimeter of the timber plate, and it is equal to the length of the wood-wood connection. Refine mesh property is applied to these areas, together with the exterior areas, and it is shown in Figure 2.45c. The FE CAE model generated by the CAD-to-FE data exchange algorithm results in uniform mesh size with a consistent aspect ratio. This helps the numerical model to avoid generating small meshes, which, in particular, increase the size of the time step. Furthermore, the mesh pattern generated by the algorithm follows the seeds defined around the perimeter of timber shell elements.

The surface elements, including the mesh data, are assembled, and the corresponding connections are then established. To do so, the node located at the middle of the length of a connection region is selected as the master node, and the remaining nodes resulting from the mesh are tied to the master node. The typical connection used for the through-tenon joints and simulated in Abaqus is shown in Figure 2.45d. For each timber plate element, the algorithm automatically collects the regions associated with the wood-wood connections, together with the coordinates of the master node per each region as a list format. Furthermore, the index of two neighboring surfaces is stored in two separate yet paired lists. The information is then used to establish the two-node link elements and define the corresponding local coordinate system. In detail, using the master node per each joint region and the paired list of connected surface elements, the two-node link element is constructed, and the stiffness of each subspring element (components of the K_{IMA} matrix in Eq. 2.34) is assigned. It is worth noting two things: The same meshing technique explained in the previous step is used in this step and since the dovetail wood-wood connections are considered rigid, the algorithm automatically joins the shared edges of the cross longitudinal and cross transverse plates within a box, which can also help to improve mesh consistency. As the last step, the entire elements are assembled to construct the CAE FE model of the entire geometry (see Figure 2.45e).

The CAD-to-FE considerably reduces the time required to prepare a CAE model. Furthermore, repetitive manual operations are removed from the design process. As a consequence, human errors, and floating-point errors are considerably minimized. Moreover, the proposed framework enhances the design flexibility of IATPs since multiple variants can be considered in the design process, while the associated structural performance can be evaluated with minimal computational expense. In the current study, the developed framework is successfully applied to a wide range of IATP structures ranging from subassembly to large-scale geometries. The case studies are shown in Chapter 3.

2.7.3 Macroscopic models[6]

Given the real geometry of an IATP element in Figure 2.46a1, the corresponding FE and macromodels are shown in Figure 2.46a2 and Figure 2.46a3, respectively. The proposed macromodel is formulated to be compatible with different quadrilateral geometries and load cases. It consists of boundary elements (Figure 2.46b), inner beam elements (Figure 2.46c, f), two (uniaxial) shear springs (Figure 2.46h, i), two-node link elements representing the joints (Figure 2.46e), and uniaxial springs distributed along the boundary elements (Figure 2.46d, g), which capture the tension-compression behavior of the plate. Overall, structural stability is maintained by connecting the inner and perimeter beam elements, the uniaxial springs, and the IMAs.

The boundary elements are pin-ended and free to rotate about their local x' and y' axes. However, the fiber-parallel (Figure 2.46c) and fiber-perpendicular (Figure 2.46f) inner beams have a flexural release about their local y' axis. In other words, the boundary elements (see Figure 2.46b) are free to rotate in the in-plane and out-of-plane directions. Simultaneously, the inner beams (Figure 2.46c, f) provide out-of-plane flexural resistance and are only free to rotate in the in-plane direction. It is within this context that the macromodel incorporates the mechanical properties of the inner beam elements to simulate the out-of-plane kinematic behavior and is independent of the boundary elements. To simulate the in-plane kinematic response, Eq. 2.22 is recalled, and two modeling strategies are used: (i) a pair of uniaxial springs in the fiber-parallel, and fiber-perpendicular directions are used to simulate the shear behavior $K_{Plate.Shr}$ (Figure 2.46h, i), and (ii) a set of distributed uniaxial tension-compression springs along the edges of the quadrilateral macromodel (Figure 2.46d, g) is used to simulate the combination of direct axial $K_{Plate.ax}$ and flexural $K_{Plate.flx}$ loads. The shear springs are unilateral, and they work only for in-plane behavior. In fact, it is assumed that the main kinematic of out-of-plane behavior is flexural deformation.

---

[6] The main parts of this section use the material published in A. Rezaei Rad, et al. [18] under the terms of the Creative Commons Attribution 4.0 International license (CC BY 4.0).

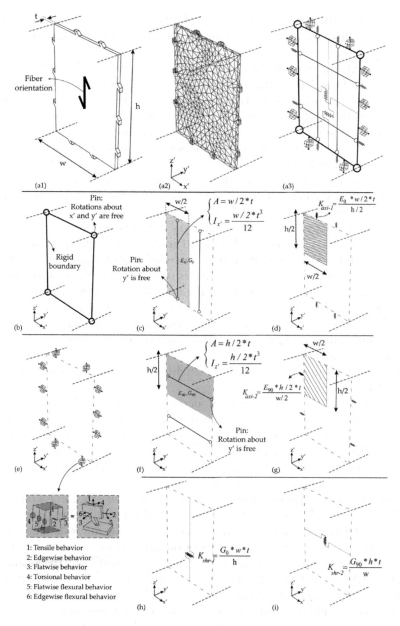

Figure 2.46 3D perspective view of (a1) the real geometry of an IATP element, (a2) an FE model, and (a3) the proposed IATP macromodel as well as (b) boundary element, (c) fiber-parallel beam, (d) uniaxial tension-compression fiber-parallel spring, (e) joints, (f) fiber-perpendicular beam, (g) uniaxial tension-compression fiber-perpendicular spring, (h) fiber-parallel IP shear spring element, and (i) fiber-perpendicular IP shear spring.

Figure credit: A. Rezaei Rad et al [18]. A. Rezaei Rad, 2020 [49]

In IATP structures, the joints and plates are one unit. In other words, because no additional connectors such as screws or fasteners are employed, the joints and the associated plate cannot be separated. Within this context, the experiments included both the joints and the neighboring region of the plate in the test specimens. With this type of experiment, the behavior of what is referred to as the 'joint region' was examined, and the initial stiffness from Table 2.1 is assigned to the two-node link elements (Figure 2.46e). Consequently, the use of the link elements to explicitly describe the behavior of the joint region can account for the local behavior of IATPs.

Given that the behavior of the joint regions is simulated using the two-node link elements (Figure 2.46e), the primary role of the boundary beams is to provide overall stability and are therefore modeled as rigid. As mentioned before, these rigid boundary elements are pin-ended so that they can freely rotate about their local in-plane and out-of-plane axes (Figure 2.46b). Moreover, by using pin-ended rigid boundary elements, the actual geometry of the timber plate is represented in the macromodel. Additionally, since the end nodes of the boundary elements have flexural releases, the behavior of the macromodel is independent of the cross section size of the boundary elements. Consequently, the inner beam elements and uniaxial shear and tension-compression springs encapsulate the mechanical properties of the macromodel. This modeling strategy makes the macromodel compatible with different polygon geometries and load cases.

To determine the cross section properties of each inner beam element, the associated tributary area, which is shown in Figure 2.46c and f, is computed. The stiffness of the uniaxial tension-compression springs located around the edges is similarly determined (see Figure 2.46d and g). The in-plane flexural stiffness of the plate is implicitly calculated using these uniaxial tension-compression springs.

Using the fabrication contours of an IATP element (Figure 2.47a), the midsurface of each 3D plate is first established (Figure 2.47a). Next, the macromodel is generated from the midsurface using the following four main steps: (1) Using the corner nodes of the midsurface to create a planar polygon (the boundary elements, Figure 2.47b); (2) identifying the location of the IMAs in the boundary elements (Figure 2.47c); (3) dividing the polygon into equal segments along the fiber-parallel and fiber-perpendicular directions and adding inner lines at the interface between segments (inner beam elements, Figure 2.47d); and (4) splitting the boundary elements at each intersection and connecting the end nodes of the inner beam elements, TT joints, and the contour corners (see Figure 2.47e).

Despite its simplicity, it is potentially possible to tune the proposed macromodel to stimulate the primary failure modes that are expected to occur in the timber plates. To do so, (i) multilinear backbone curves associated with the load-deformation/moment-rotation of the joints regions can be

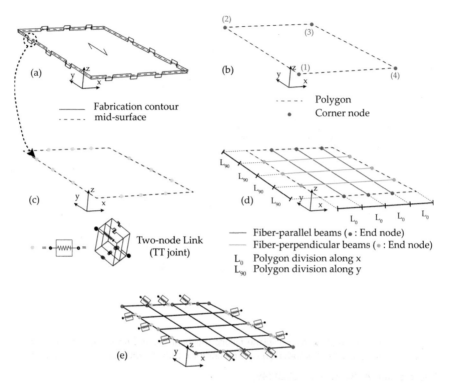

Figure 2.47 (a) 3D timber plate represented by fabrication contours and associated midsurface, (b) polygon associated with the midsurface corner nodes, (c) identifying and locating the TT joints in the boundary element, (d) dividing the polygon along the fiber-parallel and fiber-perpendicular directions and inserting inner beams, (e) establishing the boundary elements using the nodes associated with steps (b) to (d).

Figure credit: A. Rezaei Rad, et al. [18]. A. Rezaei Rad, 2020 [49]

introduced, and (ii) elastoplastic material properties, and elastic with brittle degradation properties can be assigned to the inner beams and uniaxial springs. The well-known failure mechanisms are shown in the form of crack patterns in Figure 2.48 and Figure 2.49 for the in-plane and out-of-plane loading cases, respectively. The damage mechanism associated with each crack pattern is captured with a spring or beam element of the macro-model, which is highlighted in red. Particularly, tension and compression damage [32–34] are captured by the uniaxial springs (Figure 2.48a), the well-known gross shear failure, which is also known as shear failure mode I [34–37], is captured by the uniaxial shear spring (Figure 2.48b), and the uniaxial [32, 38] and bi-axial [39, 40] flexural damage for both the fiber-parallel and fiber-perpendicular directions are captured by the inner beam elements (Figure 2.49a, b).

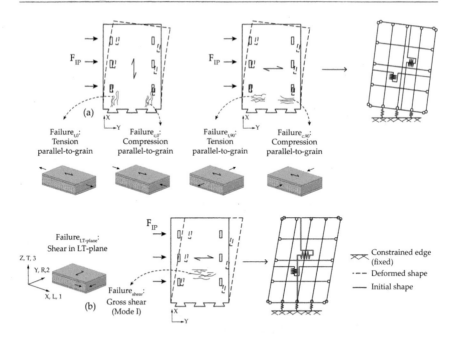

Figure 2.48 Illustrating the main in-plane failure mechanisms and how they are simulated using the macromodel: (a) tension and/or compression failure, (b) shear failure.

Figure credit: A. Rezaei Rad, 2020 [49].

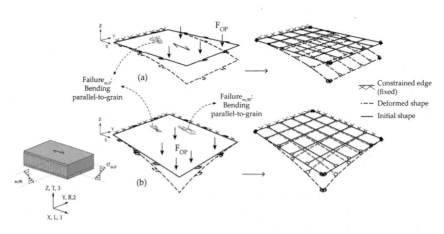

Figure 2.49 Illustrating the main out-of-plane failure mechanisms and how they are simulated using the macromodel: (a) uniaxial bending failure, (b) biaxial bending failure.

Figure credit: A. Rezaei Rad, 2020 [49].

2.7.3.1 Validation and discussions

The plate level behavior is first validated against the detailed FE models using shell elements for IP and OP loading cases. The details associated with constructing the FE model corresponding to a single IATP element were widely explained in Section 2.7.2. The macromodel, however, is constructed in OpenSees [41]. A uniaxial elastic material with infinite stiffness is assigned to the boundary elements. An elastic beam-column element is used for the inner beams based on their associated tributary area and the timber properties for each orthogonal direction. The dovetail joints are assumed rigid, while the connectors used to simulate the Through-Tenon wood-wood connections are modeled using the two-node link element. To simplify the modeling process, the same division used to define the inner beam elements is employed to determine the location of the uniaxial tension-compression springs, which are distributed around the perimeter. Consequently, these springs are placed at the end nodes of each inner beam element.

Recalling the assembly process in free-form IATP structures shown in Figure 2.5, it is recognized that the CL_i/CT_i and T_i/B_i plates are primarily subjected to in-plane and out-of-plane loads, respectively. Given this consideration, the macromodels representing the in-plane behavior of the CL_i/CT_i plates and out-of-plane behavior of T_i/B_i plates are constructed, and the load-deformation behavior is compared with the corresponding refined FE models. Regarding the macromodels, Figure 2.50 shows the section size of the inner beam elements and the stiffness values for the uniaxial tension-compression and shear springs. The inner beam elements are generated by dividing the CL_i/CT_i plates into three equal segments along the fiber-parallel and fiber-perpendicular directions. The T_i and B_i plates are divided into three and five equal segments along the fiber-parallel and fiber-perpendicular directions, respectively.

For the in-plane load case, the force corresponding to the Ultimate Limit State (ULS) is calculated using Eq. 2.35 in accordance with Eurocode 5 [17]. According to the mechanical properties of beech BauBuche described in [12-13], the shear strength of the timber boards are the weakest resisting element and control the behavior.

$$\tau_d = 1.5 \frac{V_d}{A \cdot K_{cr}} \le f_{v,d} = K_{h,v} \cdot \frac{K_{mod}}{\gamma_M} \cdot f_{v,0,k} \tag{2.35}$$

Where τ_d and $f_{v,d}$ are the applied stress and design strength values, respectively. $A = 24000 mm^2$ is the effective cross section area, V_d is the shear force, $f_{v,0,k}$ is the characteristic shear strength, $K_{h,v} = 1.0$, $K_{cr} = 1.0$, and $K_{mod} = 0.8$ are the modification factors described in accordance with Eurocode 5 [17]. The in-plane ULS load, $V_d = 176.8 KN$, is then computed and distributed along the element height at each wood-wood connection (Figure 2.51).

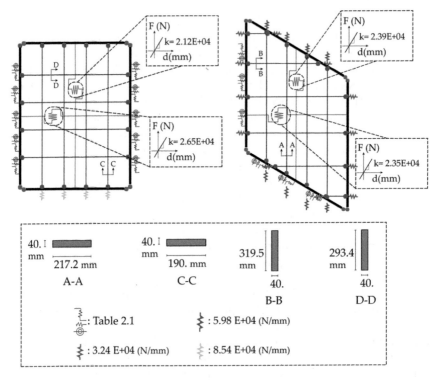

Figure 2.50 Section details and uniaxial spring properties used in the macromodel.

Figure credit: A. Rezaei Rad, et al [18]. A. Rezaei Rad, 2020 [49]

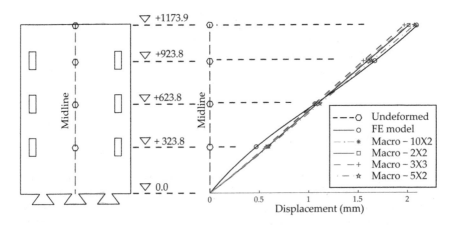

Figure 2.51 Undeformed and deformed shapes of the IATP element subjected to in-plane load (unit: millimeter).

Figure credit: A. Rezaei Rad, et al. [18]. A. Rezaei Rad, 2020 [49]

To compare the response of the macro- and FE models, the deformed shape along the element height for the different configurations defined in the sensitivity analysis are presented in Figure 2.51. The results show that the deformed shape obtained from the macromodel is similar to the one obtained from the FE model. However, the deflections in the macromodel are slightly higher in the lower portion of the element and slightly less in the upper portion of the element than that of the FE model. These differences are attributed to the different element formulations in the macro- and FE models. More specifically, the classical plate theory based on the Kirchhoff hypothesis is generally used in FE models. As such, the associated partial differential equations used to describe the displacement field are nonlinear with an order of eight. Given this consideration, the nodal degrees of freedom for each finite element is coupled in the FE formulation. This leads to differences in the curvature along the element for the FE and macromodels, where the latter uses separate (uncoupled) springs to model the shear, flexural, and axial behavior of the plate (see the off-diagonal components of the matrix K_{IMA} defined in Eq. 2.33. Accordingly, in the FE model, the rigidity of the element is distributed over the plate. In contrast, in the macromodel, the rigidity is concentrated in a specific number of discrete elements, i.e., the shear and tension-compression springs. Nonetheless, the average difference between the horizontal displacement in the FE and macromodels is 9.01%, representing an acceptable degree of accuracy.

For the out-of-plane load case, both the TT joints and timber plates play a role in the load-transfer mechanism. Generally, joints are the weakest elements in IATP structures. Specifically, the ultimate strength of wood-wood connections is lower than timber plates. Therefore, the wood-wood connections control the ULS of the out-of-plane load case. Both macro- and FE models are subjected to the out-of-plane load corresponding to the ULS of the through-tenon wood-wood connections. The undeformed and deformed shapes of the macromodels, including the configurations defined in the sensitivity analysis and the FE model, are shown in Figure 2.52. Similar to the in-plane load case, it is observed that the behavior captured by the macromodel is similar to that of the FE model. The average difference between the displacement in the FE and macromodels is 8.3%.

The time required to construct the model (data creation runtime) and run the analysis (analysis runtime) is also very interesting. As stated earlier, the joints govern the overall response of the system, and therefore, the FE model increases the computational complexity without considerably improving the efficiency. In addition to the analysis runtime, creating the model database is very time consuming for the FE models. Using only beams and spring elements, the macromodel analysis converged with no instabilities. However, depending on the element type, the adopted integration scheme, and mesh size, the FE models were susceptible to variations in the response and numerical instabilities. This is because the 1D wire elements used to simulate the wood-wood connections were connected to 2D shell elements

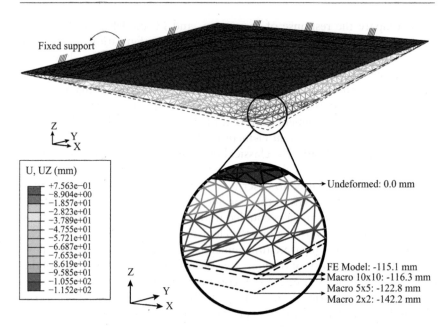

Figure 2.52 Undeformed and deformed shapes of the IATP element subjected to out-of-plane load (unit: millimeter).

Figure credit: A. Rezaei Rad, et al. [18]. A. Rezaei Rad, 2020 [49]

with high mesh density. Therefore, detailed FE simulations do not necessarily provide improvements in terms of accuracy. Furthermore, the total computational time for the FE models, including the creation of the model database and runtime, was significantly more than the time required in the macromodel. Finally, it is concluded that the joints contributed the most to the overall behavior of the system.

The compatibility of the macromodel components is maintained at the nodes. Given this consideration, and since the beam ends are located at the nodes placed along the perimeter of the plates, the modeling approach does not permit the determination of the deflection and force values on the interior of the plates. This can be improved by using multiple-segmented elements for each inner beam. Furthermore, detailed (refined) numerical models provide enhanced simulation capabilities and capture the stress-strain response and local failure modes. However, this is not reflected in the macroscopic modeling approach. Nevertheless, to better understand the local stress-strain state, it is recommended to increase the number of inner beams within the macromodel. Moreover, In the current study, the macromodel is formulated for linear elastic analysis. Prior experiments have shown that material nonlinearity can occur in the joint regions. Therefore, as a possible extension, springs with idealized multilinear backbone curves can be introduced to the macromodel to account for material nonlinearity and capture the postpeak-force behavior. Besides, the element stiffness matrix and equilibrium equations associated with the macromodel were formulated in the undeformed configuration. However, in large-span IATP structures, the effect of axial loads would be pronounced, and accordingly, geometric nonlinearity could play a significant role. As such, a possible extension could be to formulate the macromodel by incorporating geometric stiffness. More specifically, the general corotational formulation can be used to include the effect of axial forces under small-displacement compatibility relations, where the equilibrium is satisfied on the deformed configuration.

The kinematics associated with the in-plane and out-of-plane behavior of the IATP element are separately studied and formulated in this study. As such, these kinematics are uncoupled in the macromodel. Nevertheless, future studies can be put forth to understand the interaction between these two kinematics and develop sophisticated macromodels. Besides, regarding the in-plane response, the shear and axial-flexural behavior of the macromodel is uncoupled. Similar to the macromodel proposed by Kolozvari et al [42]. for reinforced concrete shear walls, an advanced macroscopic model for IATPs can be developed in the future to account for the interaction between these kinematics. Accordingly, it is recommended to incorporate timber panel behavior into a two-dimensional fiber-based model, where the coupling response is achieved at the macrofiber (panel) level. Finally, the research has shown that timber plates with quadrilateral geometries are commonly used in the parametric architectural design of IATP structures. This is mainly due to the fact that the complexity involved in the assembly of quadrilateral timber plates is low. Furthermore, the fabrication process is less time consuming than more complex geometries. Within this context, the macroscopic model proposed in this research is introduced and validated for quadrilateral timber plates. A possible extension could be to validate the macromodel for more complex geometries such as timber plates with pentagonal, hexagonal, or even octagonal shapes. As such, these macromodels can support the structural analysis and design of a more complex IATP structure, especially those that require irregular paths for the assembly of timber plates.

2.7.4 CAD-to-Macrodata transformation[7]

The design to fabrication models is not directly applicable for structural calculations software, i.e., ABAQUS, OpenSees, ANSYS. While the digital fabrication model is described by top and bottom outlines for cutting the structural models has its notations such as shell elements, beams, nodes, or other specific description defined by an engineer. The underlying mesh data structure allows keeping equal indexing logic, whereas geometry elements

---

[7] The main parts of this section use the material published in A. Rezaei Rad, et al. [29] under the terms of the Creative Commons Attribution 4.0 International license (CC BY 4.0).

such as contours, joints are described by other means of geometry types. To illustrate this, the plate's thickness is changed to a center polygon, the joints become line segments. In general, the physical properties of the material are removed and changed to low-poly geometries for a fast geometry transfer and calculation. For example, the central area of plates is defined by a separate polygon for either coarse meshing or hatching, whereas joinery zones are refined. The angle of timber joinery is removed because of the abstract model description in the engineering software pipeline. Consequently, the files are transferred by a 3D model or text files. This questions the role of the modeling process for timber plate structures: Is it is possible to create a digital model that already includes design, fabrication, and calculation data or does it require remodeling due to different interfaces coming from different stakeholders' (such as engineers, architects, and fabricators) specific methodologies? In summary, the base data structure of different models remains the same; however, the two models are nonidentical to meet specific engineering calculation standards.

The primary objective of this section is to introduce an interface to automatically convert the design CAD geometry associated with hundreds of timber plates with thousands of wood-wood connections to the corresponding macromodel. This step is not available in typical building information modeling (BIM) tools and is a major challenge in the design and performance assessment of large-scale IATP structures. Therefore, the primary objective of the current study is to formulate a CAD-to-macroalgorithmic data exchange framework that translates the CAD data associated with a custom-defined IATP structure to the associated CAE macromodel.

The parametric CAD-to-CAE exchange integrates data analysis and transformation, geometrical simplification, and structural engineering as far as timber plates with edgewise wood-wood connections are concerned. Furthermore, while the concept and methodology of the proposed CAD-to-CAE data exchange algorithm are independent of the computational platform, the current study employs (1) the Rhinoceros .NET Software Development Kit (SDK) called RhinoCommon [30] and Rhino.Python [31] to construct the plugins and components for the CAD-to-CAE data transformation, and (2) The Open System for Earthquake Engineering Simulation, OpenSees [41], to perform the structural analysis. The data exchange framework is completely designed in the CAD environment, while the output data generated by the framework is formulated such that it follows the syntax of OpenSees. The overall design workflow of the current study is shown in Figure 2.53.

Using the Common Data Model concept, the objective is to introduce an automatic framework that can incorporate a centralized repository, including the model inputs, analysis options, data management structure, CAD database, and FEM data model. The framework also enables the data exchange algorithm to parameterize the CAD and CAE information. Accordingly, the design inputs are parametrically generated and embedded

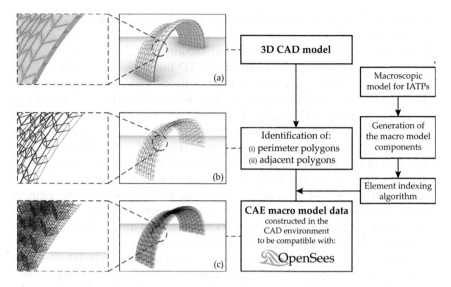

Figure 2.53 An overview of the automatic CAD-to-CAE data exchange framework for IATPs.

Figure credit: A. Rezaei Rad, et al. [29]; A. Rezaei Rad, 2020 [49]

in the code, thus enabling the design framework to update the CAD and CAE models simultaneously.

The automatic CAD-to-Macrodata transformation algorithm is centered around three design principles: (1) the intention of a design ('why' of a design), (2) the technique used to convert a CAD model to the associated CAE model according to the requirements of the CAE platform ('how' of a design), and (3) the geometric description ('what' of a design).

One of the main advantages of the CAD-to-Macrodata exchange framework is that the modeling and computational simulation in complex structures is considerably improved by using an open-source CAE platform. Furthermore, instead of working with three or more software packages, the model integrity is maintained in one platform by using object-oriented programming.

2.7.4.1 CAD-to-CAE macromodel exchange

The current investigation aims to add another layer to the design of IATPs by incorporating the macroscopic models in the structural analysis of large-scale structures. An algorithmic framework for generating the CAE macro-model from a custom-defined CAD 3D model is introduced toward this goal. The framework consists of multiple steps. First, an algorithm is introduced to compute the perimeter polygon associated with the midsurface of each quadrilateral 3D timber plate element, convert the plate from a 3D solid to a 1D polygon, and assign a unique tag to each polygon (Figure 2.53a and –b). The details of this step are presented in Section 2.7.4.2. Next, by retrieving the neighboring polygons from the mesh, 1D line elements are defined and used to represent the connection between adjacent polygons. Section 2.7.4.3 discusses the details of this step. The data associated with the polygons and connections and their ID tags are then stored as a list. In the next step, a new algorithm is formulated to generate the macromodel components from the list of data generated in the previous steps (Figure 2.53b and –c). The formulation of this algorithm is detailed in Section 2.7.4.4. Since a typical IATP structure consists of hundreds of timber plates with thousands of joints, assigning the relevant mechanical properties to each beam/spring element of the macromodel would not be possible without a systematic identification protocol. Therefore, Section 2.7.4.5 formulates an indexing algorithm to assign a unique tag to each macromodel component.

According to the proposed workflow, and given the CAD 3D geometry of an IATP structure, a data structure is designed to encapsulate the geometry of the macromodel components with the corresponding ID tags and the mechanical properties of the components. In other words, the data structure consists of the coordinate of beam and spring nodes, each of which with a unique index. The material and cross-sectional properties assigned to each beam and spring are also included in this data structure. The output provided for the CAE macromodel (Figure 2.53c) follows the syntax of OpenSees, an object-oriented open-source software for modeling and computational simulation in structural engineering. As such, the data structure formulated within the CAD environment can generate the data required for structural analysis. This, in particular, enables the user to work in the CAD modeling environment while generating the data required for a CAE platform.

The identification of the perimeter polygons and wood-wood connections and associated indexing algorithms (Figure 2.53b), as well as the automatic generation of the macromodel components and associated indexing (Figure 2.53c), are introduced to Rhinoceros 3D [43] as plug-ins using Application Programming Interfaces (APIs). While APIs within Rhinoceros 3D are available in different programming languages such as C++, C# and Visual Basic, Python, and VBScript, they use the same .NET RhinoCommon library [30]. Grasshopper 3D [44], which is a graphical algorithm editor within Rhinoceros 3D that enables visual programming, was used to edit the plug-ins and components.

2.7.4.2 CAD-to-CAE Exchange, module I: Identification of perimeter polygons and associated indexing

Identifying the perimeter polygon associated with each timber plate is the primary step in the CAD-to-CAE data exchange framework. This step is

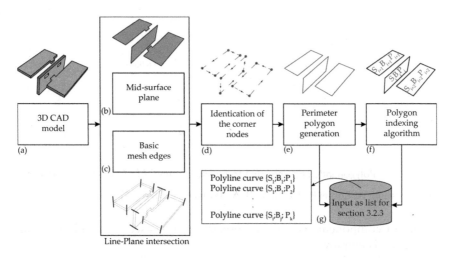

Figure 2.54 Perimeter polygon identification algorithm to be used in the CAD-CAE macromodeling.

Figure credit: A. Rezaei Rad, et al. [29]; A. Rezaei Rad, 2020 [49]

central to the CAE macromodel development because the inner beams are generated by dividing the perimeter polygons into equal segments along the fiber-parallel and fiber-perpendicular directions and adding inner lines at the segment interfaces. From the CAD 3D model (Figure 2.54a), the mid-surface plane of each timber plate is recalled (Figure 2.54b). Next, using fabrication contours available from the CAD 3D model, the basic mesh edges are retrieved (Figure 2.54c). Using the Intersection.LinePlane method available in the RhinoCommon library, the edges are intersected with the midsurface planes. The four corner nodes of the quadrilateral midsurface are identified (Figure 2.54d). Accordingly, a polygon element crossing at these four nodes is generated (Figure 2.54e). Segmented lines represent the polygon element.

The 3D CAD model was designed such that it assigns each timber plate a unique identifier index [5] to the following data components: *strips, boxes, and plates,* which are schematically shown in Figure 2.55. The most complex IATP structures are the ones with freeform geometries (Figure 2.55a). In these situations, the structure is constructed by assembling multiple strips (Figure 2.55b). Each strip is an assembly of multiple boxes (Figure 2.55c), and each box consists of four IMA-connected timber plates (Figure 2.55d). This enables the user to readily find the position of a timber plate by concatenating three numbers. Therefore, each polygon is recognized in 3D space with the following concatenated three number sequences: $S_iB_jP_k$, where S_i, B_j, and P_k, correspond to i^{th} strip, j^{th} box, and k^{th} plate. The algorithm can also accommodate simpler geometries such as beams with hollow cross sections, in which case, the first two

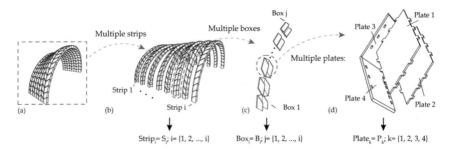

Figure 2.55 Polygon indexing algorithm to be used in the CAD-CAE macromodeling.

Figure credit: A. Rezaei Rad, et al. [29]; A. Rezaei Rad, 2020 [49]

types of components (strips and boxes) are not used, and only the plate identifier is employed.

After identifying the perimeter polygon corresponding to each timber plate, the associated identifier is retrieved from the CAD model (Figure 2.54f), and both the polygon geometry and the identifier are collected in a sorted list format using Grasshopper DataTree [45] (Figure 2.54g). The list is then used in the module described in Section 2.7.4.4 to construct the macromodel components.

2.7.4.3 CAD-to-CAE exchange, module 2: Identification of wood-wood connections and associated indexing

The second module of the CAD-to-CAE exchange uses 1D line elements to represent the connection between neighboring polygons. These lines will serve as mechanical joints in the macromodel, according to Figure 2.46g. To simplify the wood-wood connection geometry, each connection is represented by two points, one in the plate containing the tenon and the other in the plate containing the slot. To compute these points, the midsurface associated with timber plates is recalled from the CAD 3D model (Figure 2.56a, b). Also, the data associated with the neighboring plates are retrieved from the mesh. Next, a plane perpendicular to the midsurface of each plate is introduced to the middle of the tenon, and the intersection point is accordingly identified (Figure 2.56c). This point represents the tenon of the wood-wood connection. Then, by using the Line.ClosestPoint method available in the RhinoCommon library, the tenon's mate, which is located in the neighboring polygon, is identified (Figure 2.56d). Next, the line element is defined by connecting the two polygons (Figure 2.56e). This modeling technique is in accordance with the assumption made for the macroscopic model where each wood-wood connection is simulated using a two-node link.

After identifying the perimeter polygon, an indexing algorithm is introduced to assign a unique identifier to each 1D line element (Figure 2.56f). Thousands of joints exist in a typical IATP structure, and each joint has

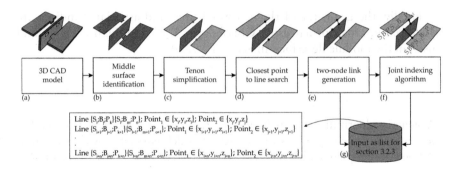

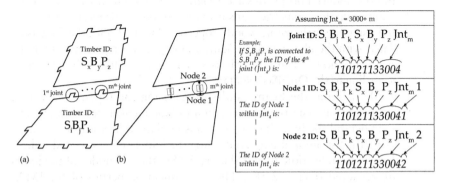

Figure 2.56 Joint identification in the macromodeling CAD-to-CAE data exchange algorithm.

Figure credit: A. Rezaei Rad, et al. [29]; A. Rezaei Rad, 2020 [49]

specific mechanical characteristics based on its geometry and fiber orientation. Therefore, assigning a unique ID to each joint is central to constructing the corresponding numerical model.

The most convenient way to assign a unique identifier number to each 1D line element is to build a new layer of information using the existing perimeter polygon IDs. Given that a joint primarily serves to connect two adjacent polygons, the indexing algorithm is formulated to include the indices of the two connected polygons. Additionally, since two neighboring polygons can be associated with more than one joint, another counter is added to the joint identifier. In other words, the indexing algorithm includes a seven number sequence where the first six identify the strip, box, and plate number of the first and second polygons, and the seventh number, Jnt_m, identifies the mth joint. Accordingly, the index of the mth joint element that connects two adjacent polygons in 3D space becomes $S_iB_jP_kS_xB_yP_zJnt_m$. Figure 2.57 shows the logic of the indexing algorithm.

Figure 2.57 Joint indexing algorithm in the macromodeling CAD-to-CAE data exchange algorithm.

Figure credit: A. Rezaei Rad, et al. [29]; A. Rezaei Rad, 2020 [49]

Finally, the geometry of the generated lines and associated indices are collected in a list format using Grasshopper DataTree (Figure 2.56g) and used in the module described in Section 2.7.4.4.

2.7.4.4 CAD-to-CAE exchange, module 3: Generation of the macromodel components

A generic algorithm, shown in Figure 2.58, is introduced in this section to generate the components of the macromodel associated with a custom-defined timber plate within an IATP structure. The main steps are as follows:

1. Using the list of input data containing the perimeter polygons and associated indices derived from Section 2.7.4.2, the four corner nodes are retrieved (Figure 2.58b in green), and two additional nodes are generated at each corner (Figure 2.58b in red). These additional nodes have the same coordinates as the corresponding corner node, and they will be used to construct zero-length rotationally free elements in the macromodel. This is in accordance with the modeling requirement of the boundary elements in the macromodel (Figure 2.46j).

2. The inner beams of the macromodel are constructed by introducing lines with equal distances along the fiber-parallel (Figure 2.58c) and fiber-perpendicular directions (Figure 2.58d). A virtual polyline connecting the four corner nodes is defined, and the length and width of each timber plate are determined. Recalling that the fiber-parallel and fiber-perpendicular directions correspond to the length and width of the timber plate, respectively, these two directions are divided into equal segments. Line elements are then added at the segment interfaces and serve as the inner beams for the macromodel, each with a unique local coordinate system. The cross-sectional properties of each inner beam element are computed based on its tributary area, and the material properties are assigned based on its fiber orientation.

3. Computing the geometric transformation of the elements (Figure 2.58e) and using the orthotropic wood properties (Figure 2.58f), the data associated with the fiber-parallel (Figure 2.58g) and fiber-perpendicular elements and nodes (Figure 2.58h) is stored in a text file with .tcl format according to the OpenSees syntax.

4. While defining the beams and springs using the OpenSees syntax, the algorithm assigns each element a unique identification index. The logic behind the indexing algorithm is explained in Section 2.7.4.5.

5. Using the list of input data containing the joints and associated indices derived from Section 2.7.4.3 (Figure 2.58i), the end nodes of the joints are retrieved (Figure 2.58j). The mechanical properties of the IMAs derived from physical experiments [7–9] are assigned to the joints (Figure 2.58m). Using the OpenSees element library, the two-node link element with six discrete springs is employed to simulate the behavior

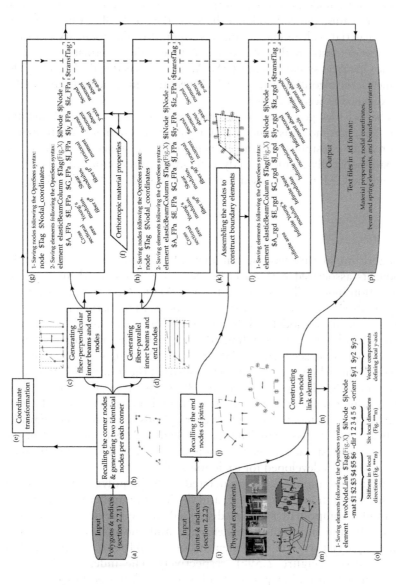

Figure 2.58 Generation of the macromodel components and data storage.

Figure credit: A. Rezaei Rad, et al. [29]; A. Rezaei Rad, 2020 [49]

of the IMAs (Figure 2.58n). The elements are stored in a text file with
.tcl format according to the OpenSees syntax (Figure 2.58o), each with
the unique identification tag explained in Section 2.7.4.3.

6. The boundary elements are generated by connecting the corner nodes, corner release nodes, end nodes of the fiber-parallel and fiber-perpendicular
inner beams, and end nodes of the joints (Figure 2.58k). A uniaxial elastic material with infinite stiffness is assigned to the boundary elements.
These elements are then stored in a text file with .tcl format according to
the OpenSees syntax (Figure 2.58l). While constructing the boundary elements, the algorithm assigns each element the unique identification index
explained in Section 2.7.4.3.

7. The .tcl files generated in the previous steps are compiled and the
CAE macromodel is generated. Given that a systematic identification
protocol is established, the elements that are subjected to loads are
parametrically defined. Analysis handlers and options are specified,
and the CAE model is used for structural analysis (Figure 2.58p).

2.7.4.5 CAD-to-CAE exchange, module 4:
Nodes and elements indexing

The algorithm in this module automatically generates the macromodel
components associated with an IATP element and assigns them a unique
identification index. This is central to the CAD-to-CAE data transformation framework because to construct the OpenSees numerical model, each
element of the macromodel and the corresponding nodes should be indexed
with an identifier. In the algorithm, (1) Crn_m is the tag that corresponds to
the mth corner node (see Figure 2.58b in green); (2) $Pcrn_m$ is the tag that corresponds to the mth corner node used to construct the rotationally free zero-
length element (see Figure 2.58b in red); (3) Trs_m is the tag that corresponds
to the mth fiber-perpendicular inner beam and its associated nodes (see
Figure 2.58c); (4) Lng_m is the tag that corresponds to the mth fiber-parallel
inner beam and its associated nodes (see Figure 2.58d); and (5) Cnt_m is
the tag that corresponds to the mth boundary element. These tags are then
linked to the ones developed for each polygon.

The first three numbers are associated with the polygon tag for each
plate, described in Section 2.7.4.2, and the fourth number represents the
macromodel component type, as discussed in the previous paragraph.
Accordingly, the index of a corner node and the nodes used for the
rotationally free zero-length element becomes $S_iB_jP_kCrn_m$ (Figure 2.59a)
and $S_iB_jP_kPcrn_m$ (Figure 2.59b), respectively. The index of the fiber-
parallel inner beam, fiber-perpendicular inner beam, and boundary element of the macromodel becomes $S_iB_jP_kLng_m$ (Figure 2.59c), $S_iB_jP_kTrs_m$
(Figure 2.59d), and $S_iB_jP_kCnt_m$ (Figure 2.59e), respectively. The
algorithm is written in the Rhino3D CAD environment using The
RhinoScriptSyntax library [46], which allows the user to create, access,

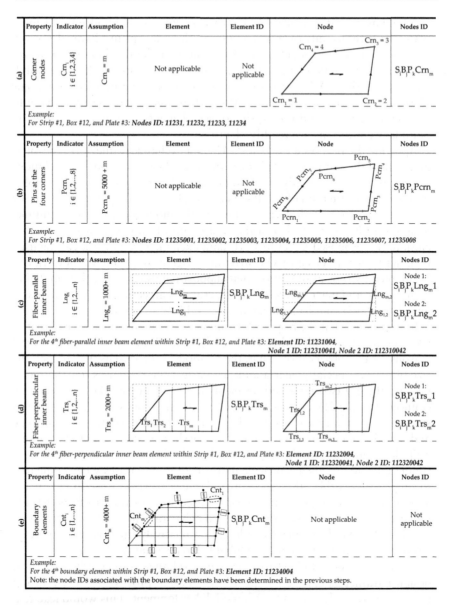

The following are the contents shown in the figure table:

(a)

Property	Indicator	Assumption	Element	Element ID	Node	Nodes ID
Corner nodes	Crn_i, $i \in \{1,2,3,4\}$	$Crn_m = m$	Not applicable	Not applicable	$Crn_4 = 4$, $Crn_3 = 3$, $Crn_1 = 1$, $Crn_2 = 2$	$S_iB_jP_kCrn_m$

Example:
For Strip #1, Box #12, and Plate #3: **Nodes ID: 11231, 11232, 11233, 11234**

(b)

Property	Indicator	Assumption	Element	Element ID	Node	Nodes ID
Pins at the four corners	$Pcrn_i$, $i \in \{1,2,...8\}$	$Pcrn_m = 5000 + m$	Not applicable	Not applicable	$Pcrn_8$, $Pcrn_7$, $Pcrn_6$, $Pcrn_5$, $Pcrn_4$, $Pcrn_1$, $Pcrn_2$, $Pcrn_3$	$S_iB_jP_kPcrn_m$

Example:
For Strip #1, Box #12, and Plate #3: **Nodes ID: 11235001, 11235002, 11235003, 11235004, 11235005, 11235006, 11235007, 11235008**

(c)

Property	Indicator	Assumption	Element	Element ID	Node	Nodes ID
Fiber-parallel inner beam	Lng_i, $i \in \{1,2,...n\}$	$Lng_m = 1000 + m$	Lng_m ... Lng_1	$S_iB_jP_kLng_m$	$Lng_{m,1}$, $Lng_{m,2}$, $Lng_{1,1}$, $Lng_{1,2}$	Node 1: $S_iB_jP_kLng_m1$ Node 2: $S_iB_jP_kLng_m2$

Example:
For the 4th fiber-parallel inner beam element within Strip #1, Box #12, and Plate #3: **Element ID: 11231004, Node 1 ID: 112310041, Node 2 ID: 112310042**

(d)

Property	Indicator	Assumption	Element	Element ID	Node	Nodes ID
Fiber-perpendicular inner beam	Trs_i, $i \in \{1,2,...n\}$	$Trs_m = 2000 + m$	Trs_1, Trs_2 ... Trs_m	$S_iB_jP_kTrs_m$	$Trs_{m,2}$, $Trs_{1,2}$, $Trs_{1,1}$, $Trs_{n,1}$	Node 1: $S_iB_jP_kTrs_m1$ Node 2: $S_iB_jP_kTrs_m2$

Example:
For the 4th fiber-perpendicular inner beam element within Strip #1, Box #12, and Plate #3: **Element ID: 11232004, Node 1 ID: 112320041, Node 2 ID: 112320042**

(e)

Property	Indicator	Assumption	Element	Element ID	Node	Nodes ID
Boundary elements	Cnt_i, $i \in \{1,...n\}$	$Cnt_m = 4000 + m$	Cnt_1, $Cnt_{m,2}$	$S_iB_jP_kCnt_m$	Not applicable	Not applicable

Example:
For the 4th boundary element within Strip #1, Box #12, and Plate #3: **Element ID: 11234004**
Note: the node IDs associated with the boundary elements have been determined in the previous steps.

Figure 2.59 Automatic indexing algorithm for the macromodel components.

Figure credit: A. Rezaei Rad, et al. [29]; A. Rezaei Rad, 2020 [49]

and manipulate a list of points, vectors, lines, surfaces, and nonuniform rational basis spline (NURBS).

Assigning a unique tag to each component of the macromodel within a timber plate has multiple advantages. First, it minimizes the number of manual operations and allows the rapid identification of each component among. Given that a typical IATP structure consists of hundreds of timber plates with thousands of joints assigning the relevant mechanical properties to each beam/spring element of the macromodel would not be possible without a systematic identification protocol. Furthermore, the compatibility between the CAD and CAE data structures is maintained in a single platform using the indexing algorithm. Moreover, after assigning the indices to the elements and nodes, the loaded elements within a structure and boundary conditions are readily identified.

Applying the proposed CAD-to-CAE data exchange framework, the interpretation of the results is straightforward when the macromodeling technique is used for structural analysis. This, in particular, facilitates seamless global performance evaluation of IATP structures. The total computational time for the macromodel, including the creation of the model database and runtime, is significantly less than that of the continuum FE model. As the main conclusion, the proposed CAD-to-CAE data exchange framework can be used by practitioners to design IATP structures. Accordingly, the framework can be adapted by typical BIM tools that support parametric modeling in digital architecture and macromodeling in structural engineering to create, manage, and share the 3D model associated with an IATP structure.

It is observed that the responses from the macromodel are closer to that of the experimental results when compared to the FE model. There are several reasons for this observation. First, the FE model is mesh sensitive. In fact, the size of mesh, especially near a joint, can affect the responses. Furthermore, because the 1D two-node link elements used to simulate the TT joints are connected to the 2D shell elements with high mesh density, the FE model is susceptible to numerical instabilities. Also, depending on the element type and the adopted integration scheme, and mesh size, the FE models were susceptible to variations in the response. It has also been shown that detailed FE simulations do not always provide more accurate responses when compared to simpler models [47, 48], especially in large-scale structures. In fact, while a large number of input parameters are used to construct a FE model, the influence of each parameter on the global behavior is not clearly understood. On the other hand, the macromodel uses a relatively small number of input parameters, and their role in the global behavior is much clearer.

With the proposed CAD-to-CAE data exchange framework, structural design parameters can be included in the design process of IATPs during the architectural design and CAD model development. This would lead to a master model that aims to optimize the design of IATP structures by satisfying the assembly and fabrication constraints and the structural engineering design requirements. Furthermore, the proposed CAD-to-CAE data exchange can facilitate design optimization of large-scale IATP structures by including the nonlinear behavior of IMAs. The nonlinear behavior, ultimate limit state, and failure criteria can be easily embedded in the spring elements of the macromodel and applied to the thousands of connections in the structure under consideration.

Although the CAD-to-CAE data exchange framework enables the automatic conversion of a CAD 3D geometry to the corresponding CAE macromodel, the information associated with the CAE macromodel is manually copied from the CAD environment and compiled in a .tcl file for further analysis in the CAE environment. To eliminate the remaining manual operations, it is recommended to conduct the CAE analysis inside the CAD environment. Finally, introducing the macroscopic modeling technique in IATPs and developing the CAD-to-CAE data exchange proposes future studies. Specifically, these advancements can lead to multiobjective design optimization, where complex geometries can be readily analyzed, and the effect of different design parameters on the global performance can be assessed within a short amount of time. Toward this end, and to parametrically define the design variables and generate the associated CAE macromodel, it is recommended to use IronPython in the future multiobjective design optimization. IronPython provides Python developers with the power of the .NET framework, and it is widely used in CAD environments.

REFERENCES

1. European Committee for Standardization (CEN), Standard EN 26891: Timber structures — Joints made with mechanical fasteners — General principles for the determination of strength and deformation characteristics, Brussels, Belgium, 1991.
2. European Committee for Standardization (CEN), Standard EN 12512: Timber structures – Test methods – Cyclic testing of joints made with mechanical fasteners, Brussels, Belgium, 2001.
3. S.N. Roche, Semi-rigid moment-resisting behavior of multiple tab-and-slot joint for freeform timber plate structures, thesis, École Polytechnique Fédérale de Lausanne (EPFL), 2017. https://doi.org/doi:10.5075/epfl-thesis-8236.
4. A.C. Nguyen, A structural design methodology for freeform timber plate structures using Wood-Wood connections, thesis, École Polytechnique Fédérale de Lausanne (EPFL), 2020. https://doi.org/10.5075/epfl-thesis-7847.
5. C. Robeller, M. Konakovic, M. Dedijer, M. Pauly, A double-layered timber plate Shell — Computational methods for assembly, prefabrication and structural design, Adv. Archit. Geom. 2016: pp. 104–122. https://doi.org/10.3218/3778-4.
6. C. Robeller, M. Konaković, M. Dedijer, M. Pauly, Y. Weinand, Double-layered timber plate shell, Int. J. Sp. Struct. 32 (2017) 160–175. https://doi.org/10.1177/0266351117742853.
7. A. Rezaei Rad, Y. Weinand, H. Burton, Experimental push-out investigation on the in-plane force-deformation behavior of integrally-attached timber through-tenon joints, Constr. Build. Mater. 215 (2019) 925–940. https://doi.org/10.1016/J.CONBUILDMAT.2019.04.156.
8. A. Rezaei Rad, H. Burton, Y. Weinand, Performance assessment of through-tenon timber joints under tension loads, Constr. Build. Mater. 207(2019) 706–721. https://doi.org/10.1016/J.CONBUILDMAT.2019.02.112.

9. A. Rezaei Rad, H. Burton, Y. Weinand, Out-of-plane (flatwise) behavior of through-tenon connections using the integral mechanical attachment technique, Constr. Build. Mater. 262 (2020). https://doi.org/10.1016/j.conbuildmat.2020.120001.

10. S. Roche, G. Mattoni, Y. Weinand, Rotational stiffness at ridges of timber folded-plate structures, Int. J. Sp. Struct. 30 (2015) 153–167. https://doi.org/10.1260/0266-3511.30.2.153.

11. S. Roche, C. Robeller, L. Humbert, Y. Weinand, On the semi-rigidity of dovetail joint for the joinery of LVL panels, Eur. J. Wood Prod 73 (2015) 667–675. https://doi.org/10.1007/s00107-015-0932-y.

12. H.J. Blaß, J. Streib, Ingenious hardwood: BauBuche beech laminated veneer lumber design assistance for drafting and calculation in accordance with Eurocode 5, Pollmeier Massivholz GmbH & Co.KG website, 2017. Retrieved from: https://www.pollmeier.com/en/downloads/design-manual.html#gref.

13. H.J. Blaß, J. Streib, BauBuche Beech laminated veneer lumber design assistance for drafting and calculation in accordance with Eurocode 5 [Baubuche Buchenfurnierschichtholz - Bemessungshilfe für Entwurf und Berechnung nach Eurocode 5], 2014.

14. Pollmeier website. About BauBuche, 2019. Available at: https://www.pollmeier.com/products/baubuche-about

15. A. Hassanieh, H.R. Valipour, M.A. Bradford, Composite connections between CLT slab and steel beam: Experiments and empirical models, J. Constr. Steel Res. 138 (2017) 823–836. https://doi.org/10.1016/J.JCSR.2017.09.002.

16. A. Hassanieh, H.R. Valipour, M.A. Bradford, Experimental and analytical behaviour of steel-timber composite connections, Constr. Build. Mater. 118 (2016) 63–75. https://doi.org/10.1016/J.CONBUILDMAT.2016.05.052.

17. European Committee for Standardisation (CEN), CEN-EN 1995-1-1:2005+A1 - Eurocode 5: Design of timber structures — part 1–1: General — Common rules and rules for buildings, Brussels, 2008.

18. A. Rezaei Rad, H.V. Burton, Y. Weinand, Macroscopic model for spatial timber plate structures with integral mechanical attachments, J. Struct. Eng. 146 (2020). https://doi.org/10.1061/(ASCE)ST.1943-541X.0002726.

19. Y. Bozorgnia, V.V. Bertero, eds., Earthquake engineering: From engineering seismology to performance-based engineering, 1st ed., CRC Press, 2004. https://doi.org/10.1201/9780203486245.

20. I.P. Christovasilis, A. Filiatrault, Numerical framework for nonlinear analysis of two-dimensional light-frame wood structures, Ing. Sismica. 30 (2013) 5–26.

21. A.C. Nguyen, P. Vestartas, Y. Weinand, Design framework for the structural analysis of free-form timber plate structures using wood-wood connections, Autom. Constr. 107 (2019) 102948. https://doi.org/10.1016/J.AUTCON.2019.102948.

22. M. Hirz, W. Dietrich, A. Gfrerrer, J. Lang, Integrated computer-aided design in automotive development: development processes, geometric fundamentals, methods of CAD, knowledge-based engineering data management, 1st ed., Springer-Verlag Berlin Heidelberg, 2013. https://doi.org/10.1007/978-3-642-11940-8.

23. A. Bagger, J. Jönsson, H. Almegaard, K.D. Hertz, W. Sobek, Plate shell structures of glass: Studies leading to guidelines for structural design, Technical University of Denmark, 2010.

24. A. Stitic, A. Nguyen, A. Rezaei Rad, Y. Weinand, Numerical simulation of the semi-rigid behaviour of integrally attached timber folded surface structures, Buildings. 2019, 9: 55. https://doi.org/10.3390/buildings9020055.

25. B.L. Gregory, M.S. Shephard, The generation of airframe finite element models using an expert system, Eng. Comput. 2 (1987) 65–77. https://doi.org/10.1007/BF01213975.

26. K. Suresh, Automating the CAD/CAE dimensional reduction process, in: Proceedings of the Eighth ACM Symposium on Solid Modeling and Applications (SM '03). Association for Computer Machinery, New York, NY, 2003: pp. 76–85. https://doi.org/10.1145/781606.781621.

27. K. Lee, Principles of CAD/CAM/CAE Systems, Addison-Wesley, 1999.

28. S.H. Lee, A CAD–CAE integration approach using feature-based multi-resolution and multi-abstraction modelling techniques, Comput. Des 37 (2005) 941–955. https://doi.org/10.1016/J.CAD.2004.09.021.

29. A. Rezaei Rad, H. Burton, N. Rogeau, P. Vestartas, Y. Weinand, A framework to automate the design of digitally-fabricated timber plate structures, Comput. Struct. 244 (2021) 106456 https://doi.org/10.1016/j.compstruc.2020.106456.

30. Robert McNeel & Associates, RhinoCommon Guides, 2020. Retrieved from: https://developer.rhino3d.com/guides/rhinocommon/

31. Robert McNeel & Associates, Rhino.Python Guides, (n.d.). Retrieved from: https://developer.rhino3d.com/guides/rhinopython/

32. L. Boccadoro, S. Zweidler, R. Steiger, A. Frangi, Calculation model to assess the structural behavior of LVL-concrete composite members with ductile notched connection, Eng. Struct. 153 (2017) 106–117. https://doi.org/10.1016/J.ENGSTRUCT.2017.10.024.

33. S.-J. Pang, G.Y. Jeong, Load sharing and weakest lamina effects on the compressive resistance of cross-laminated timber under in-plane loading, J. Wood Sci. 64 (2018) 538–550. https://doi.org/10.1007/s10086-018-1741-9.

34. D.E. Breyer, K.E. Cobeen, K.J. Fridley, D.G. Pollock, Design of Wood Structures—ASD/LRFD, 7th ed., McGraw-Hill Education, 2015.

35. M. Jeleč, H. Danielsson, V. Rajčić, E. Serrano, Experimental and numerical investigations of cross-laminated timber elements at in-plane beam loading conditions, Constr. Build. Mater. 206 (2019) 329–346. https://doi.org/10.1016/J.CONBUILDMAT.2019.02.068.

36. H. Danielsson, M. Jeleč, E. Serrano, V. Rajčić, Cross laminated timber at in-plane beam loading – Comparison of model predictions and FE-analyses, Eng. Struct. 179 (2019) 246–254. https://doi.org/10.1016/J.ENGSTRUCT.2018.10.068.

37. R. Brandner, P. Dietsch, J. Dröscher, M. Schulte-Wrede, H. Kreuzinger, M. Sieder, Cross laminated timber (CLT) diaphragms under shear: Test configuration, properties and design, Constr. Build. Mater. 147 (2017) 312–327. https://doi.org/10.1016/J.CONBUILDMAT.2017.04.153.

38. L. Boccadoro, S. Zweidler, R. Steiger, A. Frangi, Bending tests on timber-concrete composite members made of beech laminated veneer lumber with notched connection, Eng. Struct. 132 (2017) 14–28. https://doi.org/10.1016/j.engstruct.2016.11.029.

39. W.-S. Chang, J. Shanks, A. Kitamori, K. Komatsu, The structural behaviour of timber joints subjected to bi-axial bending, Earthq. Eng. Struct. Dyn. 38 (2009) 739–757. https://doi.org/10.1002/eqe.854.

40. B. Kasal, R.J. Leichti, State of the art in multiaxial phenomenological failure criteria for wood members, Prog. Struct. Eng. Mater. 7 (2005) 3–13. https://doi.org/10.1002/pse.185.

41. S. Mazzoni, F. McKenna, M. Scott, G. Fenves, OpenSees [Computer Software]: The open system for earthquake engineering simulation, 2013. Aailable at: http://opensees.berkeley.edu/.

42. K. Kolozvari, K. Orakcal, J.W. Wallace, Modeling of cyclic shear-flexure interaction in reinforced concrete structural walls. I: Theory, J. Struct. Eng. 141 (2015) 04014135. https://doi.org/10.1061/(ASCE)ST.1943-541X.0001059

43. Robert McNeel & Associates, Rhinoceros 3D website., 2018. https://www.rhino3d.com/.

44. D. Rutten, Robert McNeel and associates, Grasshopper 3D website, 2019. https://www.grasshopper3d.com/.

45. G. Piacentino, Grasshopper data trees and Python, 2019. Retrieved from: https://developer.rhino3d.com/guides/rhinopython/grasshopper-datatrees-and-python/.

46. D. Fugier, RhinoScriptSyntax, (n.d.). https://developer.rhino3d.com/guides/rhinopython/python-rhinoscriptsyntax-introduction/.

47. B. Pantò, F. Cannizzaro, S. Caddemi, I. Caliò, 3D macro-element modelling approach for seismic assessment of historical masonry churches, Adv. Eng. Softw 97 (2016) 40–59. https://doi.org/10.1016/j.advengsoft.2016.02.009.

48. I. Caliò, M. Marletta, B. Pantò, A new discrete element model for the evaluation of the seismic behaviour of unreinforced masonry buildings, Eng. Struct. 40 (2012). https://doi.org/10.1016/j.engstruct.2012.02.039.

49. A. Rezaei Rad, Mechanical characterization of Integrally-Attached Timber Plate Structures: Experimental studies and macro modeling technique. Thesis, École Polytechnique Fédérale de Lausanne (EPFL), 2020. https://infoscience.epfl.ch/record/276874?ln=en https://doi.org/10.5075/epfl-thesis-8111

Chapter 3

Case studies in Integrally-Attached Timber Plate structures: From prototypes and pavilions to large-scale buildings

Yves Weinand

This chapter describes the design-to-construction process of eight of the most recent timber plate structures primarily designed at the Laboratory for Timber Construction (IBOIS) at the Swiss Federal Institute of Technology (École Polytechnique Fédérale de Lausanne, EPFL) in Lausanne, Switzerland. The Saint-Loup Chapel in Pompaples in Switzerland, the Vidy theater pavilion in Lausanne, and the Annen headquarters in Luxembourg were designed by the author as the leading architect and structural engineer. One of the main advantages of the case studies is to extend research on the implications of advanced digital fabrication and its integration into design processes, which also connect architectural and structural design and inform each other. In this context, cooperation is a fundamental strategy for incorporating a connection design with a tectonics character. The entire process from design to fabrication is the method for an investigation into innovation and reveals the potential of wood's future when structure and architecture combine. After discussing different prototypes and recently constructed structures, the last section of this chapter introduces the main project of the current study, which is a series of freeform doubly curve double-layer at building scale.

3.1 CHAPEL SAINT-LOUP: THE FIRST FULL-SCALE FOLDED PLATE STRUCTURE

Inspired by origami, the Chapel of Saint-Loup, shown in Figure 3.1a-b and located in Pompaples, Switzerland, was the first full-scale timber folded plate structure that emerged from a technology transfer performed by IBOIS at the Swiss Federal Institute of Technology (EPFL). Given that the aim was to use this structure for a short time, it was considered a temporary system. The structure consists of 28 wall elements with 40 mm thickness and 14 roof elements with 60 mm thickness, all made from cross-laminated timber (CLT) panels. Using a reverse folding technique, the structure offers an irregular and aesthetic form with different dihedral angles between the plates ranging from 104° to 130°. The geometry of

Figure 3.1 Chapel Saint-Loup.

Photo credit: Fred Hatt and IBOIS
Project information: Chapel Saint-Loup, 2008. Client: Communauté des diaconnesses de Saint Loup. Architect: Yves Weinand, Hani Buri/Local architecture/Danilo Mondada. Timber Engineering: Bureau d'Etudes Weinand, Liège (BE). Technology transfer: EPFL IBOIS: Yves Weinand, Hans Buri.
Research credit: H.U. Buri, 2010 [1]

Figure 3.1 (Continued)

the structure was designed such that it offers a random folding logic. The assembly of miter joints between the neighboring plates was established using splice steel plate connectors with 2 mm thickness. Self-taping screws then connected the steel plates.

The form-generation tool allowed integrating structural as well as production design right from the beginning into the planning process. The chapel was entirely designed in 3D computer-aided design (CAD) software, and the geometry of the plates could be directly transferred to fabrication without modifications. Therefore, the entire project took less than six months from start to finish, and the chapel was built in less than two months. Due to the basic corrugation curve in the plan, the roof is compressed and rises up to a tip that resembles a small belfry. The result is a transition from a horizontal to vertical space, marked by the folds' rhythm. The chapel's huge wooden panels, joined by folded metal plates screwed to the timber panels, enable the structure to stay upright without a traditional linear framework. This project once again emphasized the prefabrication advantages of timber panels as well as the increasing use of computer-aided tools for generating architectural forms. Together with the information-tool technology, this substantially expanded the prospects of timber folded surface structures and led to architectural form experimentation, which enabled the overcoming of the timber panel size limitations and facilitated the realization of larger spans.

3.2 MENDRISIO PAVILION: FIRST CURVED FOLDED PLATE STRUCTURE WITH DIGITAL FABRICATION

The increase in the construction of timber plate structures is primarily attributed to folding origami art in architectural design. The *curved folded pavilion* was the first case study using prefabricated curved cross-laminated timber (CLT) panels inspired by origami folded geometries. The pavilion was presented in an exhibition at the Academy of Architecture in Mendrisio (Switzerland). The architectural design methodology was based on the curved-folding technique investigated by Buri [1] combined with integral wood-wood connections developed by Robeller [2] at IBOIS, EPFL. In fact, the increase in the popularity of wood-wood connections can be attributed to the introduction of the joinery technique in lightweight structures with irregular geometries investigated by Robeller et al. [3]. In this prototype, concave and convex curvature was employed, and the structural elements were positioned in a post manner to create mechanical stability in terms of rotation on the corners. Within this context, the design philosophy helped us to employ the concepts of the integral wood-wood connections. In light of the design philosophy developed for full-fledged structures made out of planar elements, curved timber panels were implemented in design, fabrication, and construction while satisfying structural stability. The structure had a span of 13.5 meters with CLT panels with a thickness of 77 mm. Figure 3.2a-b demonstrates the global geometry of the pavilion, together with details associated with the wood-wood connection used.

Figure 3.2 (a-b) Curved-folded thin shell structure exhibited at the Academy of Architecture in Mendrisio, Switzerland.

Photo credit: Fred Hatt and IBOIS.
Note: The geometry in (b) is regenerated based on the existing data and the framework investigated and developed by Robeller [2] and Robeller et al. [3].
Research credit: C.W.M. Robeller, 2015 [2]. Project information and technology transfer: Mendrisio Pavilion, 2013.

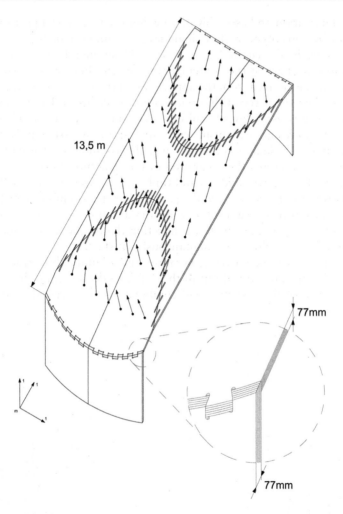

13,5 m

77mm

77mm

Figure 3.2 (Continued)

3.3 SINGLE-LAYER TIMBER FOLDED PLATE STRUCTURE WITH DOVETAIL JOINTS

Combining the algorithmic CAD and traditional carpentry, Robeller [2] at IBOIS, EPFL introduced an automatic production of wood-wood connections in thin timber folded plate structures and proposed the design-to-assembly methodology. Implementing the dovetail and Nejiri Arigate (Japanese) variants to the doubly curved timber folded plate prototype, a hexagon pattern derived from the Yoshimura origami technique was primarily researched and designed by Robeller [2], Robeller and Weinand [4, 5], and Robeller et al. [6]. The use of prefabricated timber plates was examined by the fabrication and assembly of a medium-scale physical prototype. The folded plate prototype with a single layer of timber panels using dovetail connections is shown in Figure 3.3a-b. Demountable and remountable structures are part of the contemporary design philosophy in such structures, which is derived from the application of the wood-wood integral connections. The geometry of the connections, described in Figure 3.3, allows a simultaneous assembly of multiple and nonparallel edges. Therefore, a secure interlocking mechanism with rapid and easy assembly is designed. The term 'interlocked' was used because the assembly sequence of multiple timber plates prevents the separation of the structure. In other words, the geometry of wood-wood joints and timber plates were such that there was no relative movement between the panels after the assembly. Robeller et al. [2, 6] indicates this design framework uses an all-in-one joinery technique, where the positioning of plates, connecting the neighboring elements, and load-bearing force flow are simultaneously satisfied.

To support the geometry generation and digital fabrication, numerical simulations are carried out in this book to demonstrate the structural performance of the system. Load combinations are derived according to

Figure 3.3 Folded plate structures with dovetail joints.

Photo credit: The photo in (a) is reprinted from Advances in Architectural Geometry 2014 pp 281–294 Christopher Robeller, Andrea Stitic, Paul Mayencourt et al. Interlocking Folded Plate: Integrated Mechanical Attachment for Structural Wood Panels, with permission from Springer Nature.
Note: The geometry in (b) is regenerated based on the existing data and the framework investigated and developed by Robeller [2], Robeller and Weinand [4, 5], Robeller et al. [6], and Robeller et al. [7].
Research credit: C.W.M. Robeller, 2015 [2]

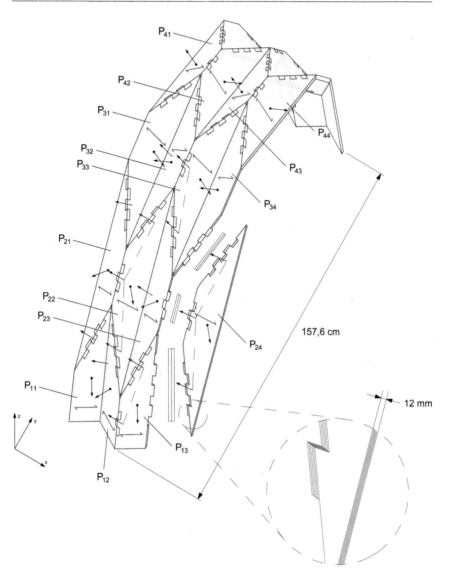

Figure 3.3 (Continued)

the Eurocode guidelines. In the current investigation, the partial coefficient for self-weight (γ_G), and the partial coefficient for variable loads (γ_Q) are considered 1.35 and 1.5, respectively. The load corresponding to the ultimate limit state (ULS) is derived 2814 N/m², 2812 N/m², and 2797 N/m² for cases where live, snow, and wind loads are dominant. The finite element (FE) analysis results indicate folded plates give a direct and intuitive perception and comprehension of geometry, and they provide a considerable amount of rigidity, as shown in Figure 3.4a-b. In other words, the

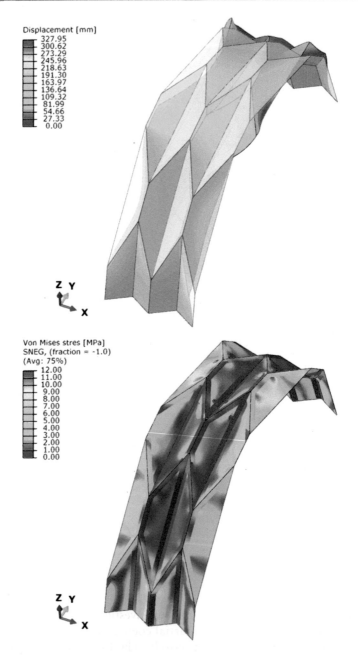

Figure 3.4 (a) Vertical displacement in mm and (b) von Mises stress in MPa associated with the single-layer folded plate structure.

Note: The geometry used in the FE analysis was generated in this study using the data in Robeller [2], Robeller and Weinand [4, 5], Robeller et al. [6], and Robeller et al. [7]

displacement field of the structure under conventional load combinations derived from Eurocode demonstrates that the structure is stiffened by a series of folded plates, although thin panels are used. Consequently, such a structural system offers a spatial structure and behaves like a reliable load-carrying mechanism. Furthermore, it is concluded that the semirigidity of the integral mechanical attachments (IMAs) should be considered in the structural design process to realistically simulate the structural performance. Moreover, at the ultimate damage state, the failure mode occurred mainly in the joint regions, where the dovetail connection was damaged and unlocked. Furthermore, the global failure of the structure was mainly controlled by the asymmetry of the double-curved shell. As the main conclusion, engineering analysis shows that the folded geometry allows the structural system to integrate a combined slab and plate action, resulting in equally efficient and elegant structures.

3.4 SNAP-FIT JOINTS: FROM A DOUBLE-LAYER FOLDED PLATE STRUCTURE TO TIMBER PEWS

Robeller [2] and Robeller et al. [7] investigated the design of timber folded plates with snap-fit joints. Figure 3.5 shows that the associated case study was a double-layer folded plate structure, where snap-fit joints were combined with dovetail joins to connect structural panels. Instead of using a single layer of thick panel, a double-layer folded structure provides greater inertia by introducing static height while not considerably increasing structural weight. Another advantage of the double-layer structure is that it enables the installation of insulation materials. As shown in Figure 3.5a-c, the interior panels first cross each other like a mortise-and-tenon joint, thanks to the snap-fit connectors, and then snap into the exterior layers above. The interior panels then double-lock the exterior panels in place, and the two additional line joints per edge improve the overall stiffness and rigidity of the connection. Given the assembly and geometry, this structural system takes advantage of the viscous elasticity of timber-derived products, and the prototype serves as a test platform of a direct interlocking of knee joints without any additional part. The physical prototype of a double-layered folded arch, presented in ACADIA 2014 by Robeller, et al. [7]. was built out of 21 mm thick panels spanning more than 2.5 meters. While the proposed prototype is a polygon in reality and not a real art structure, the structure was decomposable thanks to the distribution of snap-fit joints along with shear blocks and dovetail joints. The prototype offers a geometrically stable structure mainly due to the fact that the global form establishes a direct edgewise connection between all four layers of a fold.

A double-layer folded plate structure using shear blocks, dovetail joints, and snap-fit joints designed by Robeller [2] is shown in Figure 3.5. To support the geometry generation and digital fabrication, numerical simulations are carried out in this book to demonstrate the structural performance of the system. Load combinations are derived according to the Eurocode guidelines. In the current investigation, the partial coefficient for self-weight (γ_G) and partial coefficient for variable loads (γ_Q) are considered

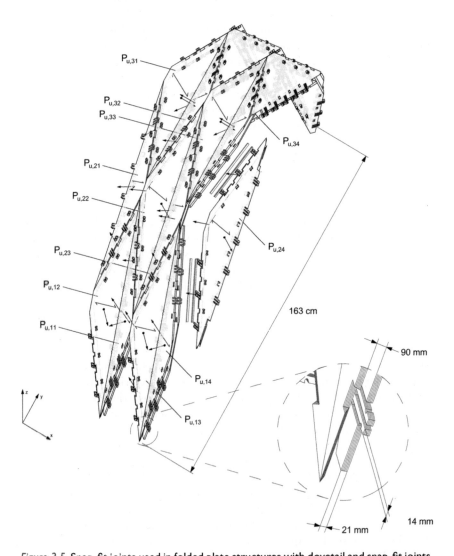

Figure 3.5 Snap-fit joints used in folded plate structures with dovetail and snap-fit joints.

Photo credit: Robeller, et al. [7]. Christopher Robeller and IBOIS.
Research credit: Robeller [2], Robeller and Weinand [4, 5], Robeller et al. [6], and Robeller et al. [7].

Figure 3.5 (Continued)

1.35 and 1.5, respectively. The load corresponding to the ULS is derived 2814 N/m², 2812 N/m², and 2797 N/m² for cases where live, snow, and wind loads are dominant. With respect to the engineering design framework introduced in Chapter 2, FE models corresponding to the prototypes were built. The results shown in Figure 3.6a-b demonstrate that the structure can maintain its functionality until the ULS. The snap-fit connections also demonstrate a considerable degree of resistance to carry shear forces parallel to the timber edges. Moreover, the assembly sequence and the geometrical form of the structure provide a full interlocking and block the relative movement between the neighboring elements. This, in particular, helps the system to carry out the tensile force and flexural moments solely through the geometry and without using additional connectors. It is concluded that the prototype allows further investigations, combining different kinds of connections and distribute them over the structural geometry. As another essential conclusion, an alternative to the ongoing construction technique is proposed by employing an on-site manual assembly without rigs, formworks, and additional fasteners.

The second application of the snap-fit joints corresponds to the design of new pews for the Cathedral of Notre Dame of Lausanne in Switzerland, where we had a range of benches and stands to be built in a very short time. The pews, built with locally sourced wood, have an innovative design that does not require glue or screws, and the associated prototype is shown in Figure 3.7. The pews are assembled in the same way described for the double-layer folded plate structure in Figure 3.5. The flexibility of the tenons

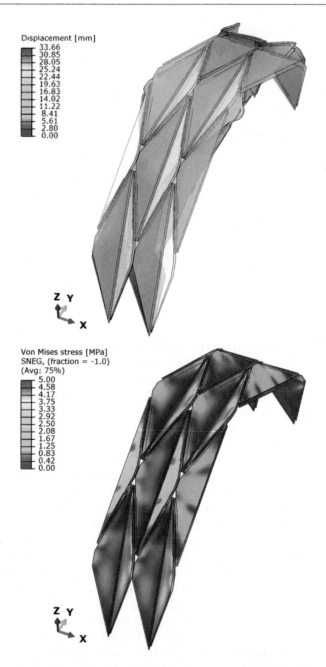

Figure 3.6 (a) Double-layer folded plate structure with snap-fit joints, (b) vertical displacement in mm, and (c) von Mises stress in MPa associated with the single-layer folded plate structure.

Note: The geometry used in the FE analysis (b-c) was generated in this study using the data in Robeller [2] and Robeller et al. [7]

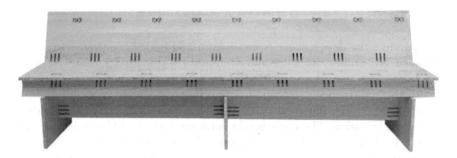

Figure 3.7 Timber pews made with snap-fit joints and installed in the Cathedral of Notre Dame of Lausanne in Switzerland.

Photo credit: Violaine Prévost and IBOIS.
Project information: Nabucco, 2018. Client: Amabilis, Renens. Architect: Yves Weinand Architectes sàrl, Lausanne. Timber engineering: Bureau d'Etudes Weinand (BE). Technological transfer IBOIS: Petras Vestartas, Martin Nakad, Loïc Pelletier.
Research credit: Robeller et al [7]., Vestartas et al. [8]

redesigned in specific geometrical terms allows for a snap-fit connection, meaning that two panels can be hold together and constructed only by snap-fit joints; this also works in reverse and makes the structure demountable. Furthermore, the entire oriented strand board (OSB) timber elements were fabricated using a computer numerical control (CNC) wood router with an accuracy of 0.01 mm. Using OSB panels was interesting because we wanted to have a very economic structure and connections.

According to the considerable amount of attention received by the project, the Lausanne Cathedral has decided to replace the rush-seated chairs that date from 1912 with the new pews using snap-fit joints. The pews, which will be placed in the Cathedral nave and transepts, are made from three-ply oak panels, are sourced from Swiss forests in Concise, Etagnières, and Cossonay in Canton of Vaud Switzerland, thus promoting local manufacture and short supply chains. They are designed to be easily dismantled and reassembled, allowing them to be moved or stored flat, thus reducing transport and storage space. Each pew, consisting of 14 planks and two wooden dowels, can be put together by two people in less than 15 minutes with a simple wooden mallet. Recalling the integration of social aspects and aesthetic aspects in digital architecture and civil engineering, the pews were fitted with reversible backs, providing added flexibility for events such as organ concerts, official ceremonies, and religious services. The backs were attached to the armrests, and a wooden peg enables them to be shifted from one side to the other in a single movement, without the use of tools. Shock absorbers prevent the backrest from banging when reversed. The seat's symmetry provides the same comfort level regardless of which side is used, and the design makes the pews easy to maintain.

3.5 SINGLE-CURVED TWO-LAYERED TIMBER SHELL

Expanding the design space from origami and folded geometries to single, low-curved arches, Robeller and Weinand [5] designed a prototype with a 3.25 meter span that consisted of two layers of timber plates equipped with through-tenon wood-wood connections as shown in Figure 3.8a-b. Nine veneer birch plywood panels with 12 mm thickness were employed for the fabrication. The static height between the top and bottom plate was set to be 48 mm, which was four times the plate thickness. In this proto-type, the Miura-Ori pattern was selected over other patterns, such as the Yoshimura type, mainly because it is geometrically compatible with roofs with low curvature. Using a quad grid, the initial double-curved surface was deformed to obtain the Miura-Ori pattern with curved element. The triangulation technique was then employed to apply the planarization of the target surface. Next, ShapeOp, an algorithm originally developed by Deuss et al. [9], was used to homogenize the dihedral angles and deter-mine the final form of the structure and the local geometry of each tim-ber plate. The prototype is shown in Figure 3.8. With respect to the force flow mechanism, the structural system offers optimized mechanical perfor-mance, especially near the wood-wood connections. In particular, the joint between plates could transfer shear and tensile forces and flexural moments

 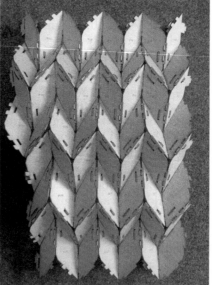

Figure 3.8 Doubly curved shell with two layers and Miura-Ori pattern [5].

Photo credit: (a) IBOIS, (b) Violaine Prévost and IBOIS.
Research credit: Robeller and Weinand [5] and C.W.M. Robeller, 2015 [2].

without any additional connectors. Furthermore, it was realized that the dihedral angle of 90° was the most suitable compromise for structural performance. Moreover, the direct connection of four plates forming a fold with two layers provides a very reliable structural system since it reduces the adverse effects of gaps and fabrication imperfections. The stiffness of the structure at the global scale is improved by a geometric and kinematic arrangement of the plates. Accordingly, it will be shown in next sections that this assembly technique is used to construct large-scale timber plate shells.

3.6 VIDY THEATRE: FIRST FULL-SCALE TIMBER FOLDED PLATE STRUCTURE WITH WOOD-WOOD CONNECTIONS ONLY

In 2017, the first full-scale of a double-layered, folded plate structure for a new hall for the Théâtre Vidy-Lausanne was realized, achieving a column-free span of up to 20 meters with a plate thickness of only 45 mm. The deployable and Origami-inspired structure was ecologically designed by the author, and it is shown in Figure 3.9a-b. This is one of the most recent applications of the digital design framework in timber plate structures. The project was also supported by the Wooden Action Plan of the Federal Office for the Environment in Switzerland. The pavilion's length and width are 28 meters by 20 meters, with a 14-meter opening scene by 11-meter depth. The height of the structure is 10.5 meters. The structure consisted of 304 different panels and 456 edge-to-edge wood-wood connections. Five-ply CLT panels with 45 mm thickness used for fabrication was manufactured in Switzerland. The irregular folded shape implied different dihedral angles with an average of 125° and a maximum of 138°. Enabled by the double-tenon connection technology, the shape of the components simultaneously serves as a joining aid for rapid and precise assembly and a direct transfer of the forces between the plates and between the two layers of the construction. This is made possible by project-specific CAD plugins. These software applications gave opportunities to customize in real-time the thickness of the layer plates, the spacing between the two layers, the type of edge connections, and the insertion direction of panels. As such, the geometry of the structure was generated in an algorithmic tool through the Application Programming Interface (API) of the Rhinoceros CAD environment. This allows an algorithmic generation of the component geometries and the computer code (G-code) for the fully automatic five-axis simultaneous machining. Given the cross-section of the structure, the structure is designed through an entirely integral attachment of four plates according to the following assembly sequence. First, the lower plate intersects both counterparts with double through-tenon connections. Next, the upper plate is introduced and intersected with the lower plate. The major benefit of the assembly technique is that the lower layer is directly attached to the upper layer without transferring the forces through additional connectors such as screws or nails. Afterward, the upper plate is inserted with a double-tenon, and the lower plate is inserted inversely. Another major benefit of the current assembly logic is that the assembly of elements guarantees a full interlocking among the entire components. In other words, entire elements of the design space are firmly held in place by other parts.

Engineering analysis was carried out using FE analysis. Similar to the previous cases, load combinations are derived according to the Eurocode guidelines. In the current investigation, the partial coefficient for self-weight (γ_G) and the partial coefficient for variable loads (γ_Q) are considered 1.35 and 1.5, respectively. The load corresponding to the ULS is derived 2950 N/m^2, 3025 N/m^2, and 2614 N/m^2 for cases where live, snow, and wind loads are dominant. Using the geometry shown in Figure 3.9b, FE analysis

Figure 3.9 A new pavilion for Vidy Theatre.

Photo credit: Ilka Kramer.
Note: The geometry in (b) is regenerated based on the existing data and the framework investigated and developed by Gamerro et al [10]., Robeller et al. [11], and Baudriller et al. [12]
Project information: Vidy Pavilion, 2017. Client: Theatre of Vidy-Lausanne. Architect: Yves Weinand Architectes sàrl, Lausanne, locally assisted by Atelier Cube SA Lausanne. Timber engineering: Bureau d'Etudes Weinand (BE). Technological Transfer: IBOIS: Christopher Robeller, Julien Gamerro, Yves Weinand.
Research credit: Gamerro et al [10]., Robeller et al. [11], and Baudriller et al. [12]

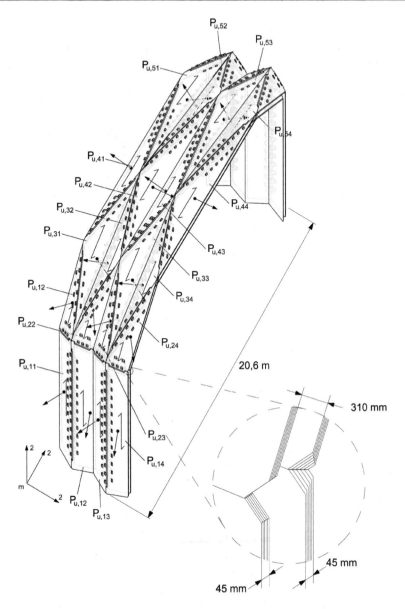

Figure 3.9 (Continued)

was conducted. The results shown in Figure 3.10a-b indicate the structure demonstrates a satisfying performance under the operational and ultimate load cases. Thus, the structure not only brings aesthetic advantages, rapid assembly, and cost savings, but it also allows the use of thin plates in irregular geometries and large-span structures.

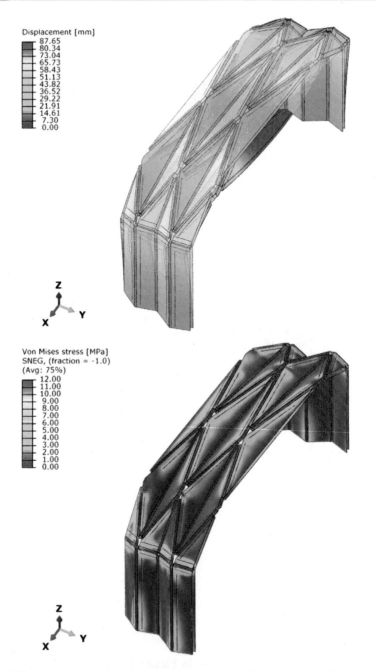

Figure 3.10 (a) Vertical displacement in mm and (b) von Mises stress in MPa associated with the single-layer folded plate structure.

Note: The geometry used in the FE analysis was generated in this study using the data in Gamerro et al. [10], Robeller et al. [11], and Baudriller et al. [12]

3.7 NABUCCO OPERA IN CATHEDRAL OF NOTRE DAME OF LAUSANNE: A SEGMENTED SPIRAL WITH INTERCONNECTED TIMBER ELEMENTS

During the Nabucco Opera in 2018, and in addition to the snap-fit wooden benches, the technological developments in wood-wood joinery were extended to an innovative geometry. In particular, an irregular segmented five-meter spiral using interconnected timber plates was presented. As shown in Figure 3.11, the scale of the spiral was chosen by the minimal bounding area of the central vault. The spiral geometry is derived from a cone shape, whose central axis and associated angle are used to position stairs within a user-specified angle and stair width. The structure was composed of 29 timber boxes made out of three-ply CLT spruce. Each box consists of nine plates: two connecting plates, two sides, one diagonal plate flipped to the triangulation of step for better structural performance, and four triangular elements. These plates are connected with integral wood-wood connections. The boxes are then screwed together and suspended by

Figure 3.11 Segmented spiral with interconnected timber elements.

Photo credit: Violaine Prévost and IBOIS. Project information: Nabucco, 2018. Client: Amabilis, Renens. Architect: Yves Weinand Architectes sàrl, Lausanne. Timber engineering: Bureau d'Etudes Weinand (BE). Technological transfer IBOIS: Petras Vestartas, Martin Nakad, Loïc Pelletier, Yves Weinand.
Research credit: Vestartas et al [8].

cables. Above the spiral, the cables are fastened to the cathedral roof, which stands 30 meters higher than the top box. The individual technical challenges, including digital fabrication and the integral joinery of plates, are consolidated step by step into a coherent framework to gain insight into the relationship between form, topology, and connections. This consolidation was made possible by successfully integrating the findings into algorithms, which allowed to develop complex procedures in incremental steps. Within this context, different design modules, namely CAD modeling workflow, prototyping, FE structural calculation, assembly, and on-site loading test, were examined within the design process.

Initially, a low-poly mesh model is used for fast design iterations. The automation of geometry generation was then carried out in Rhinoceros 3D [13]. The nesting of oriented plated was performed using an open-source OpenNest Grasshopper add-on [14]. The custom G-code tool selects geometrical elements such as lines drilling, outline for cutting, text for engraving. Furthermore, the 3D model contains the information associated with the adjacency of boxes. This, in particular, gives information about the location of the through-tenon joint, screws and holes, and the position of dowels. Within the CAD geometry generation, thickness of panels, assembly sequence, joinery (finger-joints and screws), labeling, orientation to CNC machining space, and G-code tool path generation are all created and these data are sent to the CNC machine. A 4.5-axis CNC machine was used for cutting panel outlines, drilling holes for screws and dowels, engraving panel indices, and cable inlets. Furthermore, the G-code tool path was automated from design to fabrication within the same software interface. The assembly process was carried out on-site. While the fabrication process took only two weeks, the assembly of 263 timber plates took three days. Regarding engineering analysis and design, a custom FE model was made to understand the structural behavior of the spiral. The principal objective of the calculation was to understand the rigidity of the connections and the interactions between the timber boxes. The stress in the metallic elements, as well as the cables, were compared to the ultimate and serviceability limit states (SLS). The maximum displacement and stress of the structures indicated that the structural system is reliable under SLS and ULS loads.

3.8 PREFABRICATED TIMBER PLATES WITH STANDARD GEOMETRY

This research aims to develop a circular construction system for the standardized architectural application by offering standardized structural timber elements using the wood-wood connection as primary basic building components, as shown in Figure 3.12a-k. Putting the focus on developing standard prefabricated structural elements, i.e., roof or slab structural elements, the aim is to bring these techniques into common practice, as shown in Figure 3.13a-e. Oriented Strand Board (OSB) panels are primarily used

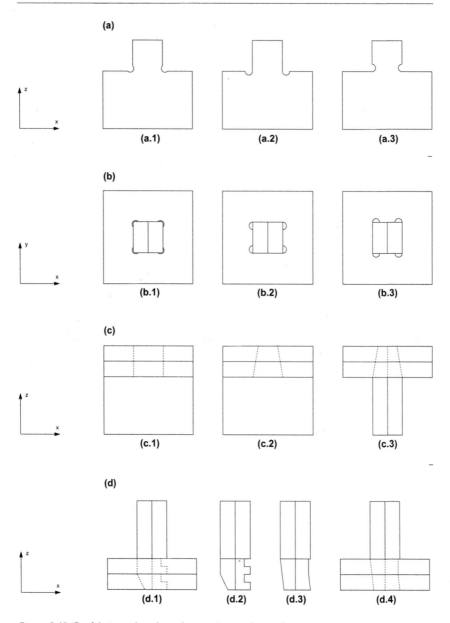

Figure 3.12 Prefabricated timber plate with wood-wood connections.

Project information: Robotic Hall Mobic, Harzé (Belgium). Location: Harzé, Belgium. Building Type: Industrial hall. Timber Company: Mobic SA (Belgium). Robotic Company: IMAX PRO (Belgium). Architect: Yves Weinand, Liège (Belgium). Timber Engineering: Bureau d'Etudes Weinand, Liège (Belgium). Technology Transfer: Laboratory for Timber Construction, IBOIS, EPFL, Prof. Yves Weinand, Julien Gamerro, Martin Nakad, Nicolas Rogeau.
Research credit: Gamerro [15] and Gamerro, et al. [16, 17].

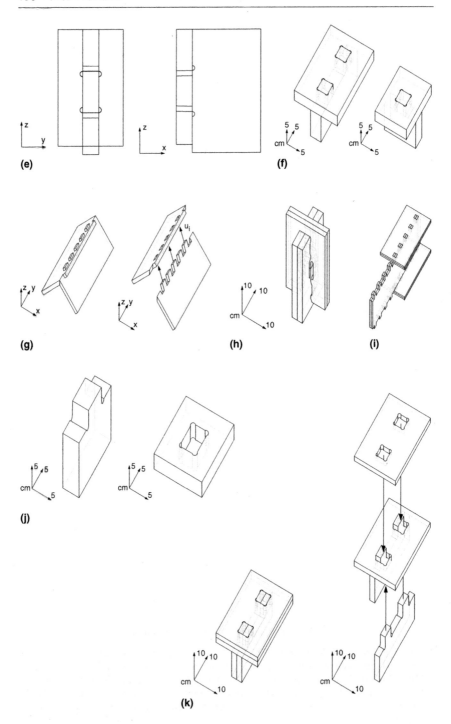

(e)

(f)

(g)

(h)

(i)

(j)

(k)

Figure 3.12 (Continued)

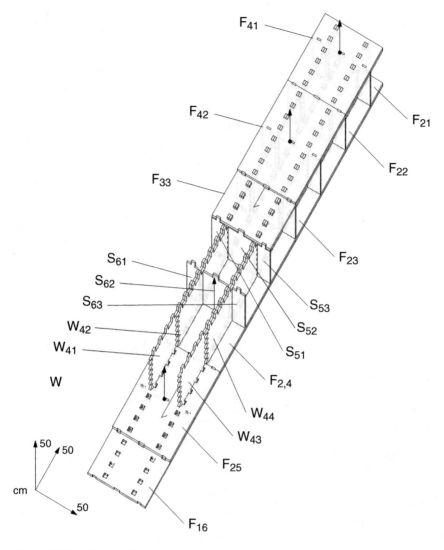

Figure 3.13 Prefabricated timber plate with standard geometry.

Photo credit: Cédric Moutschen. Project information: Robotic Hall Mobic, Harzé (Belgium). Building Type: Industrial hall. Timber Company: Mobic SA (Belgium). Robotic Company: IMAX PRO (Belgium). Architect: Yves Weinand, Liège (Belgium). Timber Engineering: Bureau d'Etudes Weinand, Liège (Belgium). Technology Transfer: Laboratory for Timber Construction, IBOIS, EPFL, Prof. Yves Weinand, Julien Gamerro, Martin Nakad, Nicolas Rogeau.
Research credit: Gamerro [15] and Gamerro, et al. [16, 17].

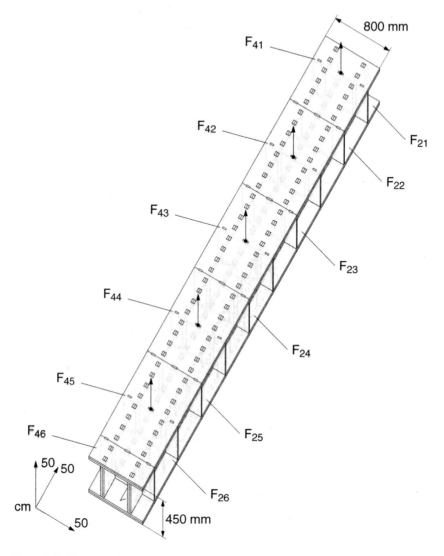

Figure 3.13 (Continued)

as the basis of the construction material. These panels are essentially cost-effective; thus, a range of I-beams was produced and tested using a physical model. In fact, OSB panels are very common, and their connections are usually made with the help of screws or nails. Using the geometry shown in Figure 3.12, roof structures or a wall structure made out of orthotropic panels is designed. Nevertheless, in general, all sorts of geometries could be performed. The fabrication constraints of such slab elements are calibrated

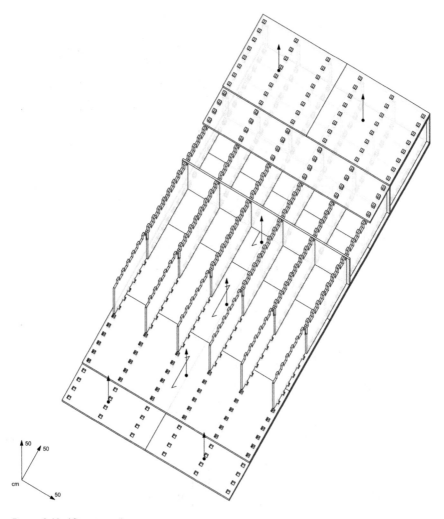

Figure 3.13 (Continued)

by commercial dimensions of the OSB panels. The basic idea still stays in the fact to create a completely demountable structure that can be reused. In addition, here, the primary structural system and secondary structural system fuse to one single surface stature. Shed roof elements and wall elements can be directly built out of those structural elements, making them very competitive.

Similar to the previous cases, load combinations are derived according to the Eurocode guidelines. In the current investigation, the partial coefficient for self-weight (γ_G) and the partial coefficient for variable loads (γ_Q)

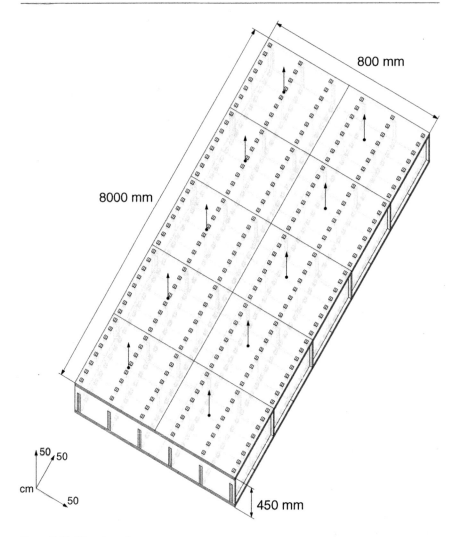

Figure 3.13 (Continued)

are considered 1.35 and 1.5, respectively. The load corresponding to the ULS is derived 2905 N/m², 2775 N/m², and 2732 N/m² for cases where live, snow, and wind loads are dominant. Regarding the engineering analysis discussed in Chapter 2, FE models corresponding to the prototype shown in Figure 3.13 were built. The results, shown in Figure 3.14a-b, indicating that the wood-wood connections are mechanically efficient even when OSB panels are used, and they satisfy the current standards for timber engineering, especially Eurocode 5 [18].

Figure 3.13 (Continued)

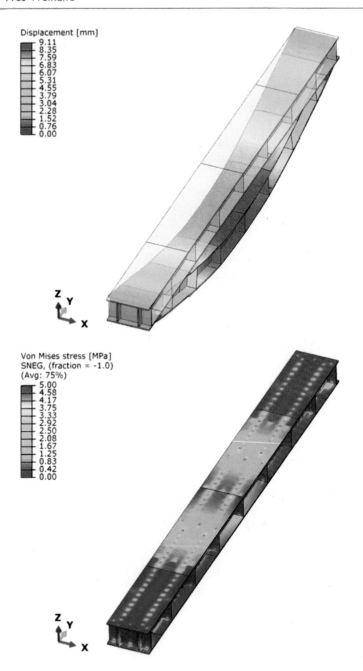

Figure 3.14 (a) Vertical displacement in mm, and (b) von Mises stress in MPa associated with prefabricated timber plate with standard geometry.

Note: The geometry used in the FE analysis was generated in this study using the data from Gamerro [15], and Gamerro et al. [16, 17]

3.9 DOUBLE-LAYER FREEFORM TIMBER PLATE STRUCTURE

After reviewing and demonstrating the different case studies in this chapter, the most complex Integrally-Attached Timber Plate (IATP) structure designed by the architects Yves Weinand/Valentiny and subject to a major technology transfer at IBOIS EPFL is presented herein. The project called Annen SA head office [19] is the most recent structure that consists of all design elements and features explained in Chapter 2, and it pushes the boundaries of construction. The entire project, located in Manternach, Luxembourg, consists of a series of 23 discontinuous double-layered double-curved timber plate shells that will accommodate a 5800 m² facility for timber prefabrication. The target surface of the 23 arches is shown in Figure 3.15a. Each arch is a double-curved shell structure with a design inspired by Eladio Dieste's Gaussian masonry vaults, and the vaults present overlapping s-shaped cross-sections. The span of the arches ranges from 22.5 meters to 53.7 meters, and the height and width of each arch are nine meters and six meters, respectively. The southern part of the project is a single-story factory space that narrows down to the northern end of the site. In the shorter direction, the shells follow an S-shape to have a second curvature to prevent buckling (see Figure 3.15b). The arch design is inspired by the Gaussian Vaults of the Uruguayan architect and engineer Eladio Dieste, as used in the TEM factory in Montevideo, Uruguay (1960–1962), and the Caese Produce Market in Porto Alegre, Brazil. The shells' span-to-rise ratio varies from 2.5 to 6 regarding the catenary line, which Dieste used for the masonry roofs (with a ratio varying from 8 to 10), mostly in the first four meters from the ground plane. Here the arches are tangential to the vertical axis to reduce horizontal forces at the supports. The construction method is based on timber plate components (boxes), each composed of four plates. The geometry requires multiscalar modeling techniques, including nonuniform rational basis spline (NURBS) subdivision into mesh, fabrication, and structural calculation data shown in Figure 3.15c for architectural representation, CNC cutting, analysis of the structure, and validation of the assembly sequence. The unique interpretation of Eladio Dieste is based on segmented timber plate elements interconnected using the wood-wood connection to form a continuous shell structure.

Each arch has its own exclusive freeform geometry design using custom-developed CAD plugins. Since the geometry of this project corresponds to a freeform shape, the project represents the most complex form of IATP structures. A general description of the geometry, timber plate elements, and associated assembly logic, and the one degree of freedom (1DOF) wood-wood connections are provided in Chapter 2, Section 2. To provide a consistent design-to-construction plan for this project, years of research and technology transfer are spent, a wide range of prototypes have been designed and constructed, and several physical experiments have been carried out.

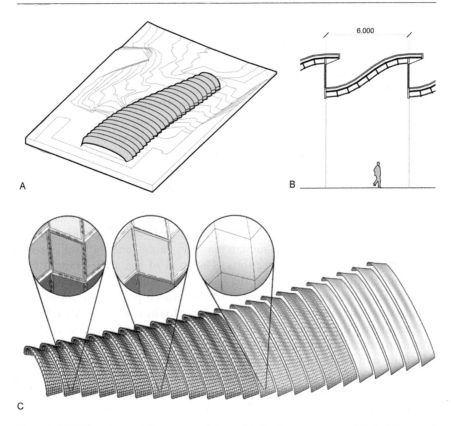

A

B

6.000

C

Figure 3.15 Different model representations: (a) Design geometry, (b) Architectural detail, (c) Models representing design, fabrication and structural calculation (right-to-left).

Project information: Annen Project, ongoing. Location: Manternach, Luxembourg. Client: Annen Plus SA. Architect: Yves Weinand, Liège/Valentiny hvp architects, Remerschen. Timber Engineering: Bureau d'Études Weinand, Liège (Belgium). Technology Transfer: Laboratory for Timber Construction, IBOIS, EPFL, Prof. Yves Weinand, Didier Callot, Petras Vestartas, Dr. Anh Chi Nguyen, Dr. Aryan Rezaei Rad, and Dr. Christopher Robeller (NCCR Digital Fabrication funded contributors).
Research credit: A.C. Nguyen, 2020 [25]; A. Rezaei Rad, 2020 [26]; C. Robeller, et al. [20].

The two-layer structure of 40 cm thickness timber plates is constructed from quad mesh. The plate geometry follows both the graph structure of the underlying mesh and its geometrical properties such as mesh-face normals, edge lines, and vertices. A list of parameters allows changing the rotation of plates, plate thickness, offset between layers, as well as special cases for boundary elements. The system is composed of modules. Each module consists of four plates: two side elements connected using dovetail joints and a top and bottom plate connected using tenon-mortise connections (see

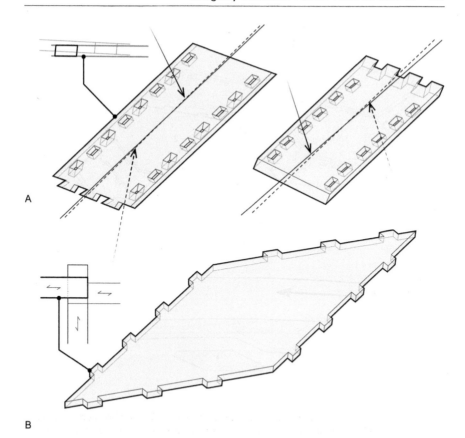

A

B

Figure 3.16 The insertion plane varies from neighbor to neighbor due to the overall curvature of the mesh (a). The top and bottom plates (b) are inserted into side plates using the tenon-mortise joint.

Figure 3.16a-b). Due to the thickness of the timber plates, a reciprocal pattern is introduced. It helps to rotate plates around each vertex node resulting in a small gap allowing to assemble boxes without collision.

There are no plates for the remaining vertical faces of a quad because these plates are shared with the neighboring segments. While each joint is represented as a pairwise connection joining two elements, the side wall employs a three-valence joint that groups two pairwise connections. This method allows reducing material use of one-third compared to the case when all four edges are treated as a timber plate instead of two. The tenon-mortise joint design must consider the three-valence connection by enlarging the distance between tenons. The intersection area of the three plates is divided into alternating segments, creating slots that receive the tabs of the shell plates. This allows for direct contact between the shell plates for the transfer of compressive forces between these slots – the vertical plate holds the shell plates and the top and

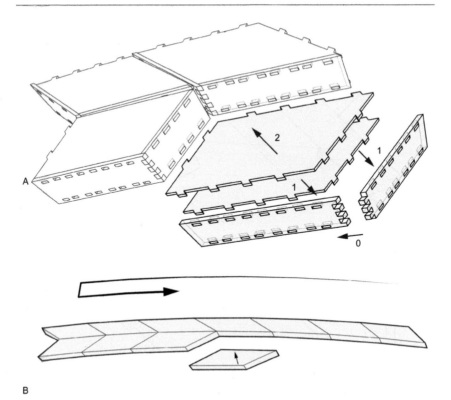

Figure 3.17 The shell is constructed from box components (a). The box components are made from four timber plates: two side plates and top and bottom elements. The insertion order of box components follows linear zig-zag pattern (b) to block each element and previous row of connected boxes.

Research credit: Robeller, et al. [20, 21].

bottom face of all tabs are held in palce. The joinery process, together with the insertion order of box components, is shown in Figure 3.17a-b.

With respect to the fabrication, the algorithm for CNC fabrication is based on G-code generation from the top and bottom outlines to perform five-axis cutting. The method considers tool radius and notches needed to cut concave corners of inner and outer contours. The arches were fabricated from 4 cm LVL panels. The first shell is composed of 1248 elements requiring 1063 m^2 of LVL. The shell weighs 32 tons and is 43 m^3 in volume. The fabrication table, together with some IATP timber plates, is schematically shown in Figure 3.18a-c.

3.9.1 Preliminary design phase

The design of freeform double-layer timber plate structures started with the initial physical prototype shown in Figure 3.19. The initial design was

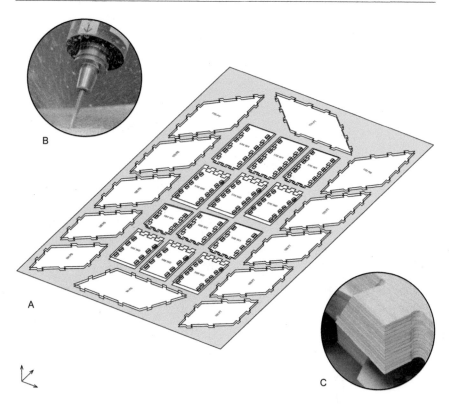

Figure 3.18 Timber plates from the 3D model have to be oriented to 2D space and nested to 40 mm LVL panels packed as tightly as possible to minimize waste (c). The fabrication technique is a five-axis CNC machine (a) needed to cut changing angles of panels, including digital machining artifacts such as Mickey Mouse ears (b) at the concave corners of joints.

essential to understand the assembly patterns. As a secondary objective, the design of timber plates, aimed to avoid doubling all surfaces within a box-beam structure. Within the geometry exploration stage, two alternative ideas were taken into account: (1) the structure consists of several closed boxes, where each box includes six timber plates connected by means of wood-wood connections. In this case, each box exists independently and it is produced as a discrete element; and (2) at least one side of each box is left open to enable the sequential assembly of multiple boxes without additional connectors. In this context, only after the assembly process, the boxes are closed. This prototype shows that the second idea offers a straightforward solution by connecting all boxes with surfaces left open at first and closing during the assembly sequence. As a consequence, fabrication constraints directly determine the geometrical and thus mechanical quality of the wood-wood connection.

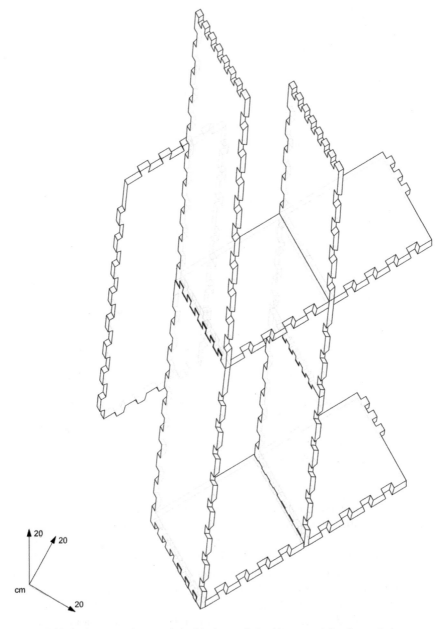

Figure 3.19 Preliminary design of double-layered double-curved freeform timber pate structures.

3.9.2 Assembly and fabrication explorations at component level

To gain insight into the fabrication process of wood-wood connections, the prototype shown in Figure 3.20a-b was investigated. The design methodology was applied at a 1:1 scale using 40 mm thick LVL panels made from beech hardwood. The hardness of the material itself and the diameter of the cutting tool have direct consequences on the assembly process, and they determine the geometry of tenons, associated net surface, and local geometrical singularities. Furthermore, the prototype served to verify the complexity involved in assembly at the scale of conventional buildings. The force that needs to be applied to integrate those panels was achieved with the use of cranes or human strength. Within the fabrication and assembly process, it was observed that even geometrical imperfections do not lead to considerable errors. The transfer from the digital workflow toward the physical realization gives the necessary credibility of the system and allows us to travel further down the path of innovation.

The assembly of the segments follows their numbering. First, a box component is assembled by connecting side walls by dovetail joints and then inserting the top and bottom plates. Second, the box components are connected one by one in a linear manner after the indexing of each row. The shells are assembled lying on the longer side, inserting the box component with a hammer. The components are connected with through-tenon joints using one insertion vector per box. The individual shells are connected with shear plates joined with metal fasteners on site. The whole shell is assembled, lying on the long edge. When the structure is assembled, it is lifted in place by a crane.

3.9.3 Assembly and fabrication explorations at structural level with scale 1:3.33

The exhibition at the Advanced Architectural Geometry conference 2016 hosted at ETH Zurich allowed us to collaborate on a small-scale prototype for the first time. The industry partner collaborated with us to assemble the prototype. The prototype with a 7.1-meter span, 1.75-meter width, and two meter height, shown in Figure 3.21, was essentially fabricated and assembled at EPFL Switzerland with the Annen SA company's help [19]. We used 15 mm-thick spruce plywood panels as the construction material, and similar assembly logic described in Chapter 2 was employed. The montage sequence was verified using timber panels with a scale of 1:2.7. This offered a fast, precise, and simple assembly, allowing for constructing a series of differently shaped shells without a costly mold or support structure. Instead, inclined joints cut with a five-axis CNC milling machine embed the correct location and angle between plates into the shape of the parts. This constrains the relative motions between joined parts to one assembly

Figure 3.20 Assembly and fabrication explorations at the component level for double-layered double-curved freeform timber plate structures.

Photo credit: IBOIS, École Polytechnique Fédérale de Lausanne (EPFL), Switzerland
Research credit: A.C. Nguyen, 2020 [25]; A. Rezaei Rad, 2020 [26]; C. Robeller, et al. [20].

Figure 3.21 Assembly and fabrication explorations at the prototype level for double-layered double-curved freeform timber pate structures.

Photo credit: Aryan Rezaei Rad. Research credit: C. Robeller, et al. [20]

path. To take advantage of the benefits of such connectors, the constrained assembly paths must be considered in the fundamental design of the system, allowing for the insertion of each plate. The fabrication and assembly process engaged in the prototype also provided an overview and estimate of the construction time necessary to mount the real structure. The same prototype was then demonstrated in Zurich, which helped the company become familiar with the assembly process and construction planning process. Moreover, the design and construction of the prototype helped to gain confidence in the coding and digital production process.

3.9.4 Structural explorations at structural level with scale 1:1

To verify the design methodology and gain insight into the performance of IATP structures, a medium-scale prototype was designed, fabricated, and assembled (Figure 3.22a–e). Accordingly, experimental tests were carried out. The cutting process was optimized, and all connections were more precisely executed. The fabrication process could thus be validated by producing this prototype. In addition, mechanical research was performed on this prototype. This prototype was realized after the initial prototypes made for assembly and fabrication explorations signifying that for such complex

Figure 3.22 1:1 subassembly level associated with double-layered double-curved free-form timber pate structures.

Photo credit: A.C. Nguyen, 2020 [25]

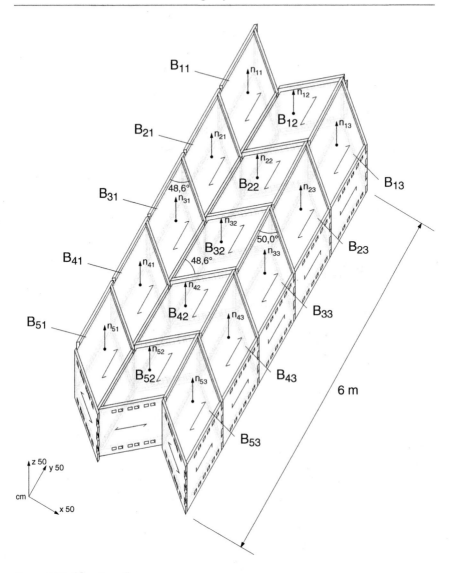

Figure 3.22 (Continued)

spatial structures, fabrication and geometrical observations must first be established before moving on to the mechanical investigations.

To verify the design methodology explained in Chapter 2, the performance of a medium-scale prototype was first investigated through a physical experiment. Extracting the geometry from a large-scale prototype with a 24-meter span with a scale of 1:1, the prototype consists of 15 boxes arranged in the format of 5×3 boxes, each with the length, width, and static height of 1.17 meters, 0.85 meters, and 0.6 meters, respectively. The

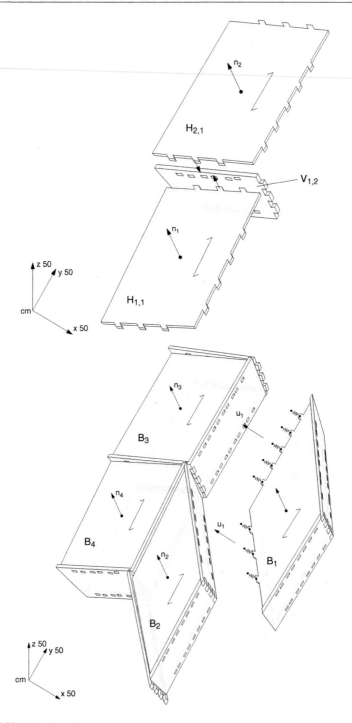

Figure 3.22 (Continued)

arrangement of boxes forms a slab-type structure with a length of 6.85 meters and a width of 2.59 meters. Within this configuration, each box is connected to at least two other boxes. Consequently, the global geometry leads to a minimum side effect caused by the boundary supports. Given that each box consists of four plates, the prototype had 60 timber plates, each with 12 wood-wood connections located around its perimeter. The wood-wood connections have the same geometrical dimension used in the small-scale experimental tests explained in Chapter 2. The geometry associated with the prototype is shown in Figure 3.23. In order to study the replicate-to-replicate variability, three specimens of the prototype were fabricated and assembled at EPFL, in Lausanne, Switzerland.

The experimental investigation starts with designing the test setup. To do so, the structure was first mounted on and fixed to HEA 160 steel profiles. Next, the system was mounted on two support rigs at its two ends, as shown in Figure 3.23. This allows the structure to freely rotate about the global axes X, Y, and Z. Specifically, one support is hinged, which restrains the structure from translational movements (see Figure 3.24a),

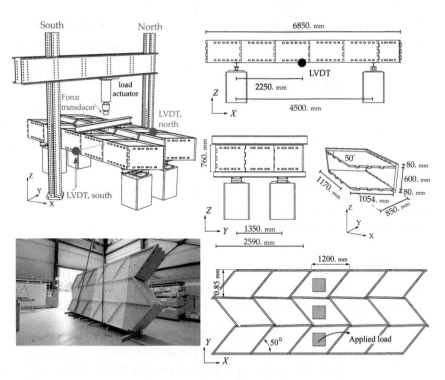

Figure 3.23 Three-point bending test setup for the 5x3 box prototype, the dimension of the boxes and timber plates, and the measurement instrumentation.

Photo credit: Anh Chi Nguyen and IBOIS.
Research credit: A.C. Nguyen, 2020 [25]

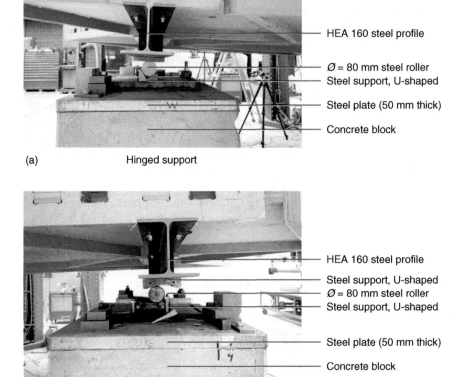

(a)   Hinged support

(b)   Rolled support

Figure 3.24 Boundary conditions in the prototype: (a) hinged support, (b) rolled support.

Photo credit: Anh Chi Nguyen and IBOIS.
Research credit: A.C. Nguyen, 2020 [25].

and the other support is rolled, which restrains the system only from translations along the Y and Z axes (see Figure 3.24b). The supports were then
attached to the four concrete blocks shown in Figure 3.23. Next, a three-
point bending test was employed by subjecting the prototype to a static
load conforming to the European Standard EN 26891 [23]. The load is
applied to the three top panels located at the midspan of the structure using
a hydraulic jack with 200 kN capacity, as shown in Figure 3.23. An HEB
240 steel profile is attached to the hydraulic jack to uniformly apply the
load to the plates. Two Linear Variable Differential Transducers (LVDTs)
were installed on the north and south sides of the physical specimen to measure the vertical deformations, and a force transducer was attached to the

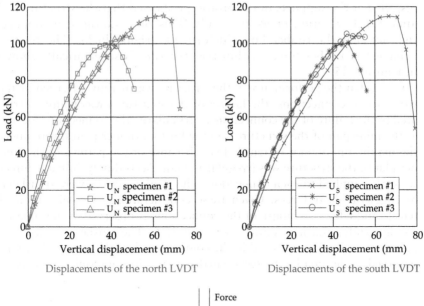

Displacements of the north LVDT   Displacements of the south LVDT

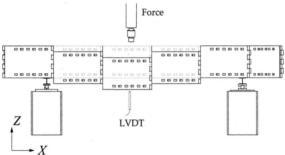

Figure 3.25 (a) Load-displacement curves obtained at midspan with LVDTs installed in the north (in blue) and south (in red) of the prototype. (b) The schematic performance of the prototype and relative deformation between the boxes.

Research credit: Nguyen et al [22]. And A.C. Nguyen, 2020 [25].

load actuator (hydraulic jack). The details associated with the installation of the measurement devices are shown in Figure 3.23.

The load-displacement behavior associated with the LVDTs installed in the north and south side of the midspan of the structure is shown in Figure 3.25. The average maximum load at the failure damage state recorded by the force transducer was 107.23 kN with a coefficient of variation of 0.07. The low dispersion in the force-deformation response indicates that the test setup was optimally designed. During the test procedure, it was observed that the timber boxes deform like rigid blocks. This, in particular, indicates that wood-wood connections play a more important role than the timber plates. Moreover, the inclinometers installed at the supports of the prototype showed that the structure had almost 1.3° rotation at its two ends. This observation also supports the fact that the timber boxes deform like rigid blocks. The relative behavior of timber boxes within the structure is schematically shown in Figure 3.25.

Given that the performance of the structure was mainly governed by the wood-wood connections, the failure of the structures was mainly attributed to the failure of the connections. Particularly, the connections located at the mid-span of the structure and the bottom timber plates were under the maximum amount of tensile load. Because of the pattern of the timber plates, the structure was irregular in plan. Accordingly, in-plane forces were also applied to each connection. The in-plane forces were translated to edgewise shear forces. Given these considerations, a combination of tensile and edgewise failure of the connections is observed at the collapse damage state, and it is shown in Figure 3.26. The tensile and edgewise failure evidence that occurred in the connections had the same indications observed in the small-scale tests. Furthermore, since multiple connections

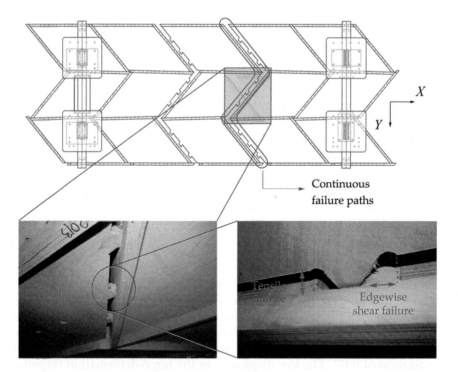

Figure 3.26 Failure of the prototype appeared in a continuous failure path at the bottom of the prototype and because of a combination of shear and tensile loads.

Photo credit: Aryan Rezaei Rad and IBOIS.
Research credit: Nguyen et al [22] and A.C. Nguyen, 2020 [25].

failed under the load case, they formed a continuous failure line. This is highlighted in Figure 3.26. As noted earlier, the failure pattern depends directly on the herringbone pattern used in the design of the structure. This pattern caused a continuous failure path and separation between the boxes. It will be shown in Chapter 4 that by changing the design pattern, the failure mode can be controlled, and an optimized system can be used in the design process.

Similar to the previous cases, load combinations are derived according to the Eurocode guidelines. In the current investigation, the partial coefficient for self-weight (γ_G) and the partial coefficient for variable loads (γ_Q) are considered 1.35 and 1.5, respectively. The load corresponding to the ULS is derived 3810 N/m², 3800 N/m², and 3797 N/m² for cases where live, snow, and wind loads are dominant. Numerical results, shown in Figure 3.27a-d, are compared to a three-point bending test performed on two specimens. The developed spring model shows promising results for its application to a full double-layered timber plate shell. However, the spring model is stiffer than the linear regression of the two tested specimens.

3.9.5 Annen Max 1 prototype

The final scale of the panels was decided to be 40 mm thick. Furthermore, the scale of the arch was 1:2. We determined it was essential to test the montage assembly process as well as the montage of each arch in this module. Working on this perspective, it became clear that we had to mount each range of the arch horizontally, one by one. Given that final boxes of the large span arches can have a length of up to 2.4 meters, it becomes difficult or even impossible to move them up and then insert them into the system. The prototype is shown in Figure 3.28a-b.

Using the same load combination, the engineering analysis was carried out in the ABAQUS CAE environment, and the behavior of the structure was explored. The results indicate that the performance of the structure satisfies the objectives dictated by the design standards, especially Eurocode 5. Furthermore, the force that appeared in the wood-wood connections satisfies the design target of the Eurocode standard as well, and they remain in their serviceability limit state. The displacement of the structure along the vertical direction and the von Mises stress are shown in Figure 3.29a-b.

3.9.6 Annen Max 2 prototype

Annen Max 2 is the first full-scale prototype of the Annen project. The construction of this project allowed for a detailed check of all production and construction constraints; together with fabrication planning, CNC commanded automatic labeling, storage, and assembly. Intermediate storing of all panels and the full construction of the arch is also addressed in this phase of the project. As described in Section 3.9.3, the montage of

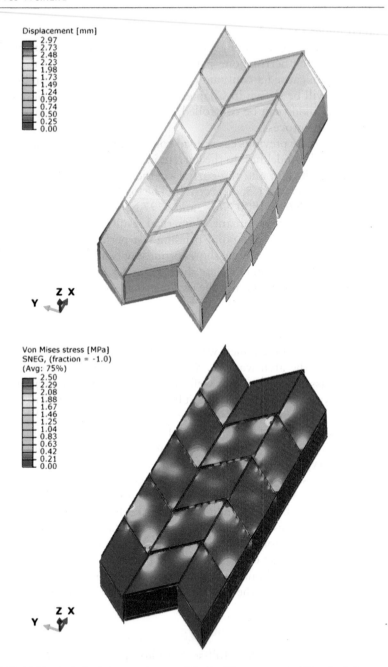

Figure 3.27 (a) Vertical displacement in mm, (b) von Mises stress in MPa, (c-d) forces and stresses in wood-wood connections associated with double-layered double-curved freeform timber plate structures.

Research credit: Nguyen et al [22] and A.C. Nguyen, 2020 [25].

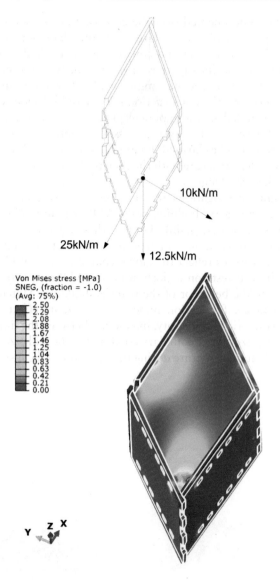

Figure 3.27 (Continued)

each arch requires a proper and specific montage sequence. But the remaining question is to see how this sequence has been realized manually and in what specific geometrical position each arch needs to be assembled by single manpower. In this regard, the author's architecture and engineering firm, Bureau d'Etudes Weinand, spent several working sessions to discuss montage joints and their consequences of the assembly process with the contractor. Even though the predominant assembly sequence had been discussed and experimented over the constructions of several prototypes, the erection process of such a large structure had not been addressed. In this regard, the planning and construction team had to deal with different challenges. Specifically, they have to identify the position where the arch is assembled. Also, the maximum weight of each box should be determined. Furthermore, the construction team should decide whether subensembles are included in the montage process and, if so, which size can be tolerated for them. Moreover, a possible montage joint between those subensembles should be clarified. Finally, identifying the kinematics needed to link the wooden structure with the prepared concrete foundation system was another key challenge.

Experimental investigations remain necessary to gain insight into the structural performance of building-scale IATP structures. Physical tests seem inevitable in such scales, mainly because they capture those features that cannot be detected when only a portion of the structure is physically tested. Such exclusive features reflect the global characteristics of IATP structures. Therefore, physical tests on a portion of an IATP structure cannot represent or simulate the behavior of the real-size building. Material variability, assembly tolerance, fabrication imperfection and its effect on the global structures, geometric nonlinearity, interaction between the timber plates in a freeform shape, size effect, boundary condition of the real-size structure, and loads imposed on the structure during the assembly and construction are the

Figure 3.28 Annen Max I prototype.

Photo credit: IBOIS, École Polytechnique Fédérale de Lausanne (EPFL), Switzerland. Research credit: A.C. Nguyen, 2020 [25]; A. Rezaei Rad, 2020 [26]; C. Robeller, et al. [20].

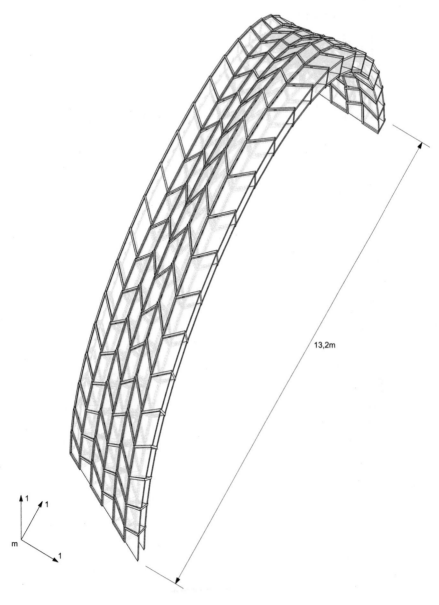

Figure 3.28 (Continued)

parameters that are reflected only in large-scale experimental tests and have a considerable influence on the mechanical behavior of the global structure. A large-scale nondestructive experimental test is conducted on a full-scale arch constructed by the Annen Plus SA head office project [19].

In the current investigation, two experimental tests are conducted. In the first full-scale test, the performance of arch #22 is individually studied. This

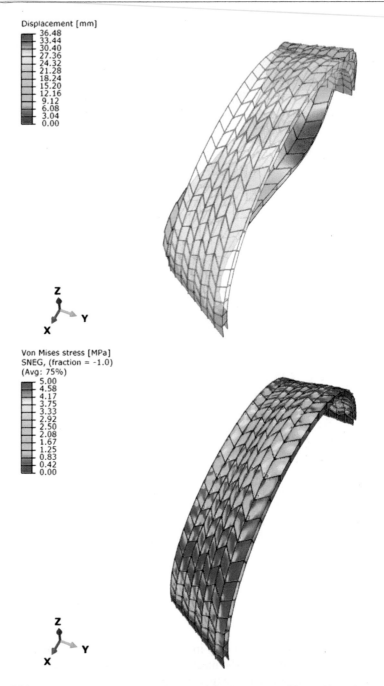

Figure 3.29 Annen Max I prototype: (a) vertical displacements, (b) von Mises Stress.

* Research credit: A.C. Nguyen, 2020 [25] and A. Rezaei Rad, 2020 [26].

arch consists of 792 planar timber plates, each with multiple tenons or slots located around its perimeter. These timber plates form 200 four-sided timber boxes. In the second test, the behavior of arch #22 and half of arch #21 is researched. The goal of the second test was to understand the interaction between two consecutive arches and verify the functionality of the supports of the structure. A row of timber caps is attached to each arch enabling the connection to the neighboring arches with multiple 100 mm ×100 mm × 5mm steel struts. Figure 3.30a-b shows the overall geometry associated with the arches. Details associated with the several design steps of the project will be discussed in the following sections.

Typical support designed for the arches is shown in Figure 3.31. Given the geometry of arch #22, eight steel plates with 15 mm thickness, together with M16 bolts, are used to connect each arch to the 25 mm thick steel plates that are attached to the foundation. With this type of boundary condition, it is guaranteed that fix-end support for the structure is provided. The attachment between these steel plates and the concrete foundation is provided by M27 tie rods anchored using grouting mortar.

The timber plates used to construct arch #22 and half of arch #21, which were 1188 elements in total, were fabricated by the industry company, Annen SA [19], and transported to the construction site. For this purpose, five-axis CNC machines using tools with a diameter of 25 mm were employed for the digital fabrication process. The G-code required for the fabrication process is generated through a loft-like 3D offset of each timber plate contour polygon.

Inserting the tight-fitting timber plates one by one and according to the assembly plan, the four-sided boxes are first manually assembled on the ground. The assembly of boxes then follows the assembly process. Once the entire arch is constructed, a crane is used to lift it all at once and position it in its determined location. The process of transporting an entire arch is shown in Figure 3.32. The structure is then fixed to the supports according to the boundary conditions described in Figure 3.31.

Project information: Annen Project, ongoing. Client: Annen Plus SA. Architect: Yves Weinand, Liège/Valentiny hvp architects, Remerschen. Timber Engineering: Bureau d'Études Weinand, Liège (Belgium). Technology Transfer: IBOIS, EPFL, Prof. Yves Weinand, Didier Callot, Petras Vestartas, Dr. Anh Chi Nguyen, Dr. Aryan Rezaei Rad, and Dr. Christopher Robeller (NCCR Digital Fabrication funded contributors).

Reiterating the assumption made earlier, two types of experiments are carried out for the large-scale structure. In the first one (hereafter: the first test case), the performance of arch #22 is researched, where two load cases are considered. In the first load case, the structural performance under its self-weight is investigated. In the second load case, the performance of the arch under the applied loads is investigated. For the second experimental test (hereafter: the second test case), the behavior of arch #22 and half of arch #21 is studied. The same load cases used in the first

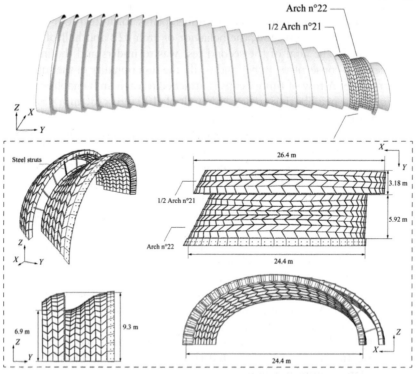

Figure 3.30 Ilustration of the Annen Plus SA head office project in Manternach, Luxembourg, consisting of 23 double-layered double-curved timber plate shells, and the experimental tests carried out on arch #22 and half of #21.

Photo credit: IBOIS, École Polytechnique Fédérale de Lausanne (EPFL), Switzerland.
Research credit: A.C. Nguyen, 2020 [25]; A. Rezaei Rad, 2020 [26]; C. Robeller, et al. [20].

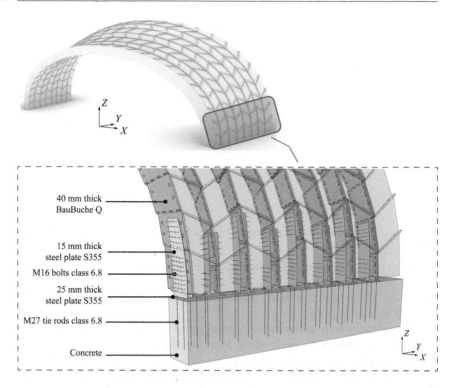

Figure 3.31 Typical boundary conditions used for the timber plate freeform arches.

Research credit: A.C. Nguyen, 2020 [25].

Figure 3.32 Assembly and transportation of an arch to its determined location, Annen SA, Luxembourg.

Photo credit: Peter Zock, Annen.

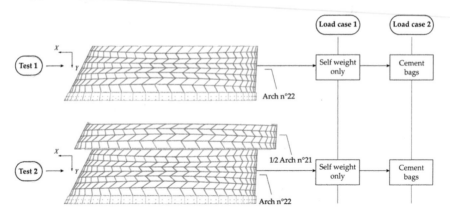

Figure 3.33 Two test cases and associated two load cases used in the large-scale experimental test.

Research credit: Nguyen et al [24] and A.C. Nguyen, 2020 [25].

test case are adopted herein. The test and load cases are schematically shown in Figure 3.33.

The primary step in the experimental setup design is to adopt an appropriate methodology to apply static loads to the structure. While different loading protocols can be potentially applied to large-scale surface structures, in the current investigation, the primary focus is put on understanding the critical behavior of the large-scale prototype under vertical loads. In detail, given that the primary responsibility of spatial structures is to transmit surface loads such as self-weight, snow, and wind to its supports, it is necessary to design a test setup that can apply uniformly distributed loads to the structure. Through gravity, surface loads can be applied to the spatial structure using different approaches such as water bags, sand bags, sealed air bags, gas pressure systems, and discrete loads with suspended dead weights. However, given the size and scale of the structure, and considering the fact that the test should be carried out on-site, the most appropriate and feasible approach should be adopted. The discrete loading approach was not feasible among the possible approaches, mainly because it was not a cost-effective approach. This approach was also quite sensitive to wind-load effects, and it could potentially apply additional dynamic loads to the structure because of its movement during wind time. After assessing the alternative approaches, it was determined that the distributed surface loading process using cement bags was the best candidate. The advantages of this method are attributed to its flexibility and adaptation to changes in the geometry of the structure. Furthermore, this method guarantees a high degree of safety. Moreover, variations in the amount of the applied load are minimum when the structure deforms. Also, because of the arch's low curvature, the cement bags were retained in their position without additional components.

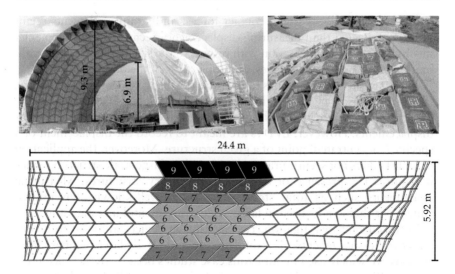

Figure 3.34 Loading the prototypes using cement bags applied on top of arch #22: The loaded area together with the number of bags are highlighted.

Photo credit: Anh Chi Nguyen.
Research credit: Nguyen et al [24] and A.C. Nguyen, 2020 [25].

The structure was loaded using 25 kg cement bags distributed over the top surface. Computing the worst-case scenario regarding the vertical displacements at its midspan, the arch is partially loaded. Figure 3.34 shows the top view of arch #22 and the timber boxes that were loaded by the cement bags. The number of the cement bags used for each box is also shown in Figure 3.34. A total load of 55 kN was applied to the structure. Based on the area loaded by the cement bags, the average surface load was 1.5 kN/m². This amount of load is used to ensure that the structure stays at its linear elastic phase. The cement bags located on each top plate of a box component were distributed in a way that they provide a uniform distributed load and avoid any load concertation. A crane was used to put each cement bag one by one on top of the structure without imposing a dynamic effect.

Different intensity measures can be adopted to evaluate the performance of large-scale surface structures. However, the condition of the construction site, the complex geometry of the structure, complicated interaction between the timber plates and wood-wood connection do not allow us to record force, stress, and strain fields. Thus, structural displacement is considered as the primary performance measure in the current physical experiment. In fact, measuring displacements in large-scale physical tests is deemed a reasonable way to understand a specific structure's performance. This is mainly because well-known techniques are available to carry out the measurement. Moreover, the stiffness of the structural system can be readily computed by having the corresponding displacements in the structure. On the other hand, measuring local stress, contact forces, and strain of elements seems more complicated than measuring displacements because of the complexity involved in the instrumentation with respect to the size of the structure, calibration of the associated devices, and robustness of the outputs.

There are several techniques to measure the displacements of a structure. Electrical transformers – also known as linear variable displacement transducers (LVDTs) – are among the most efficient techniques to measure linear displacements. However, LVDTs are meant to record the displacement of a specific (concentrated) point of a given structure. Moreover, the application of LVDTs does not seem a practical approach in large-scale structures since it requires the presence of additional temporary structures. Measuring instruments such as total stations are other well-known conventional equipment that can be used in large-scale experimental investigations.

In the current investigation, two measuring devices are used to record structural displacements: (1) a total station and (2) 3D laser scanner. Total stations are widely used in numerous case studies because of their accuracy in the millimeter range. However, such devices are highly dependent on the number of target points used in the measurements. Furthermore, total stations require a low data-sampling rate. Accordingly, an unobstructed line of sight between the targets and the device is necessary. On the other hand, 3D terrestrial laser scanning is a measuring technique that has been gaining popularity. This device enables high-precision capture, measurement, and analysis of a 3D space. The laser scanner provides the user with the point cloud corresponding to the scanner object. Accordingly, the 3D coordinates of millions of points scanned by the device are collected and stored for postprocessing and determination of structural displacements. However, this technique is time consuming, and the postprocessing computation depends on the size of the point cloud. Furthermore, this technique adopts a relatively complicated calibration. The instrumentation and outputs of the total station and laser scanner devices are discussed in the next section.

In the current study, 16 targets positioned on the bottom layer of the arches are used to measure structural displacements using the electronic Leica Geosystems TCR 305 total station device (Leica Camera AG, Wetzlar, Germany). A microprismatic reflective target with a dimension of 3 mm × 3 mm × 0.5 mm is glued to an aluminum plate and attached to each station. In addition to the 16 stations, five reference stations are installed at stable locations on the concrete foundation to calibrate the total station device and guarantee that any additional movement and noise is detected, recorded, and consequently, removed in the post-processing step. According to the measurements, the accuracy of the device, reflected in standard deviation, is 5 mm for reflective targets within the range of 1.5 meters to 80 meters. The device was installed on a stable, concrete foundation with approximately 10 meters far from station #12 and along the global Y direction. The position of the targets and total station, together with the test setup used on-site, are shown in Figure 3.35.

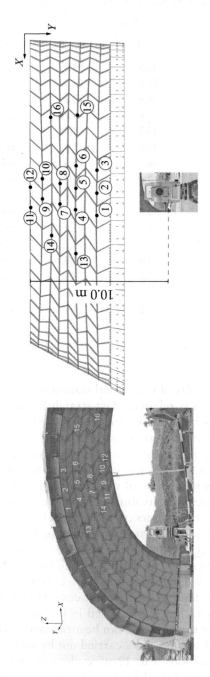

Figure 3.35 Location and notation of the stations installed on the bottom layer and the location of the total station device.
Research credit: Nguyen et al [24] and A.C. Nguyen, 2020 [25].

Table 3.1 Displacements of the 16 targets measured with the total station for test #1 and #2 [24]

Station ID	Total station		Unit
	Test 1	Test 2	
1	20.5	14.0	mm
2	22.3	15.0	mm
3	18.4	13.3	mm
4	19.8	14.3	mm
5	19.8	15.4	mm
6	15.9	11.9	mm
7	16.5	13.9	mm
8	17.5	15.2	mm
9	12.9	14.4	mm
10	12.0	13.2	mm
11	11.3	12.9	mm
12	11.1	13.7	mm
13	7.6	7.4	mm
14	5.6	9.5	mm
15	10.1	6.6	mm
16	11.8	4.9	mm

One of the major benefits of using total station is that the postprocessing of the recorded data is relatively fast straightforward. Furthermore, the entire process can be done on-site, and accordingly, new measurements can be carried out if any noise occurs during the measurements. In this project, the data postprocessing and computing the displacement of the total target stations for both test cases #1 and #2 was approximately ten minutes. A minimum amount of noise was detected during the measurement. This indicates that loading the top plates and recording the displacements of the bottom layers is an efficient experimental methodology. The vertical displacements are documented in Table 3.1. Comparing test cases #1 and #2, the results indicate that the connection between the arches (test case #2) improves the stiffness of the system. Accordingly, the average of the structural displacement in the second test is approximately ten percent lower than that of the first test case. However, the displacements of stations 9 to 12 locally increased in the second test case by 14 percent. The reason for this increase can be attributed to the fact that the self-weight of arch #21 was partially carried out by arch #22. Given this, and considering that stations 9 to 12 were close to arch #21, there was a local increase in the number of final displacements compared with the first test case.

3D laser scanners are high-speed terrestrial devices that offer 3D measurements as well as image documentation. With these devices, nondestructive tests were carried out in this project to record, measure, and postprocess the displacements associated with the physical objects using a line of laser light. Through 3D laser beam scanning, this system collects millions of points and generates the corresponding point clouds and high-resolution 3D images. FARO® Focus 150 laser scanner (FARO Technologies Inc., Lake Mary, FL, USA) is used in the current investigation. This device is able to scan 122,000 to 976,000 points per second. Furthermore, the scanner covered the horizontal and vertical field of view by 360° and 300°, respectively. Moreover, the 3D position accuracy of the scanner was 3.5 mm at 25 meters.

Calibration of the scanners and the scanned objects' registration are the primary steps when 3D laser scanners are used. The registration targets are mainly responsible for aligning multiple scans seamlessly. To do so, spherical shapes are generally used as reference stations. Using such reference shapes allows the highest possible scanning efficiency from various directions. Consequently, errors associated with axis mapping are minimized. The spherical reference shape is a hollow made of plastic with a distinctive surface to achieve excellent reflective properties. The object is supplied with a magnetic holder. Accordingly, it can be easily fixed on different surfaces.

The registration process for using 3D laser scanners is done by installing five artificial spherical targets on the concrete foundation of the structure. As shown in Figure 3.36, these spherical targets are positioned at the corners and center of arch #22. The registration starts with the geometrical transformation of the coordinates of the point clouds. In this step, each point cloud's local coordinate system is gathered and translated to a global coordinate system. Using the spherical markers, automatic registration is conducted by FARO® SCENE software (FARO Technologies Inc., Lake Mary, FL, USA).

To guarantee that the entire design space, including arches #21 and #22, are scanned and to ensure that the point cloud can be reliably used to compute the displacements, the structure is scanned from different coordinates. For the first test case, the structure (arch #22) is scanned from three different locations, while for the second test case, the structures (arch #22 and half of arch #21) is scanned from five different locations. This is because the second test case consists of a larger design domain than that of the first test case, and more stations were required to cover the space. These locations are highlighted in red and blue for test cases #1 and #2 in Figure 3.36b-c, respectively. Those areas in the point cloud that do not belong to the arches are then removed from the design space. This process is manually carried out in the Rhinoceros 3D computer-aided design (CAD) environment®, version 6.0 (Robert McNeel & Associates, Seattle, WA, USA).

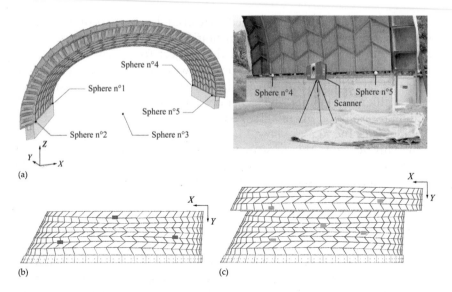

Figure 3.36 3D (a) Laser scanning setup, (b) installation of reference sphere shapes for test case #1, and (c) installation of reference sphere shapes for test case #2.

Research credit: Nguyen et al [24] and A.C. Nguyen, 2020 [25].

The 3D laser scanner rotates 3413 times to provide a full scan and the corresponding point cloud of the structure. Using a resolution of one-fifth with 4x quality, each 360° rotation collects 8192 points, and therefore, a complete scan of the structure consists of 28,400,000 points. The time spent carrying out a full scan was eight minutes and 15 seconds. Arch #21 scanned by the 3D laser scanner is shown in Figure 3.37. A 10 × 10 mm of the point cloud in a plane surface is also shown in Figure 3.37.

Similar to the case where the total station was used, the main focus of the test using the 3D laser scanner was put on documenting the displacements of the bottom layers of the arches. After postprocessing the point cloud, the initial postprocess results indicated that the spherical stations for cloud registration led to an approximately 10 mm error. It is within this context that another step of cloud registration was manually carried out to reduce the error. To do so, the *CloudCompare* software package was employed, and the final displacements derived from the point cloud database are derived. The structural vertical displacements of test cases #1 and #2 are shown in Figure 3.38.

Using the terrestrial laser scanning system, the distribution of displacements throughout the surface of the structure is readily derived and computed. Furthermore, the values obtained from the laser

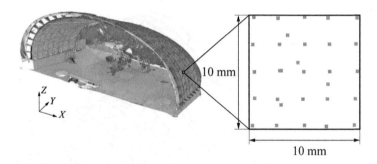

Figure 3.37 Point clouds associated with the first load case and an area of 10 mm × 10 mm extracted from the cloud.

Research credit: Nguyen et al [24] and A.C. Nguyen, 2020 [25].

scanning system are relatively 24 percent different from the values obtained from the total station. This is attributed to two different sources. First, the displacements computed in the terrestrial laser scanning system depend on the cloud-to-cloud distance computation. Second, the total station can measure the displacement of a discrete point with a high degree of accuracy. Simultaneously, the terrestrial laser scanner is a powerful tool to measure the displacement of a surface area. Moreover, the size of the structure studied in this experiment is quite large, making the computational process involved in cloud computing difficult and complicated.

Finally, and for the sake of structural design, the structure was subjected to the load combinations derived from the Eurocode. The results, including the vertical deformations and von Mises stress, are shown in Figure 3.39.

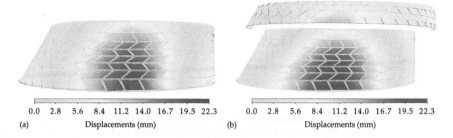

Figure 3.38 Top view of the arches: Structural vertical displacements of (a) test case #1 and (b) test case #2 using the 3D laser scanning technique FARO®.

Research credit: Nguyen et al [24] and A.C. Nguyen, 2020 [25].

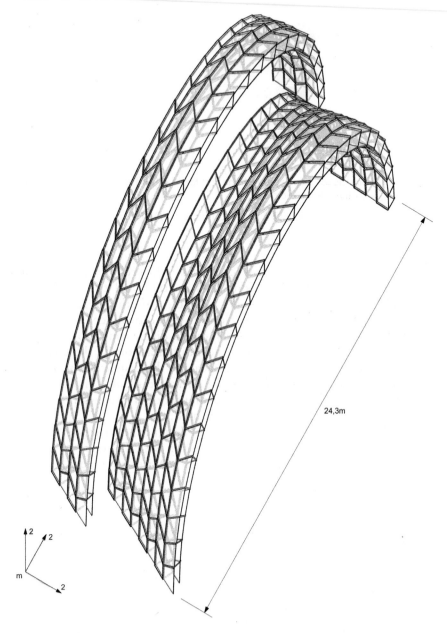

Figure 3.39 (a) 3D CAD geometry of the prototype, (b) vertical deformation in mm, (c) von Mises stress in MPa, and (d) forces in the wood-wood connections.

Research credit: Nguyen et al [24] and A.C. Nguyen, 2020 [25].
Note: The geometry and the FE model in (a-c) are regenerated based on the existing data and the framework investigated and developed by Nguyen [25] and Nguyen et al. [22]

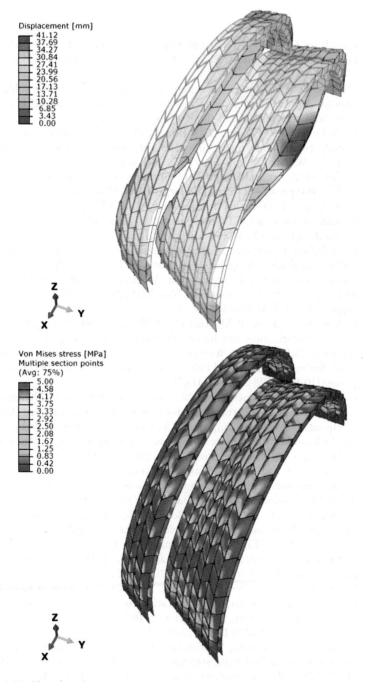

Figure 3.39 (Continued)

REFERENCES

1. H.U. Buri, Origami — Folded plate structures, thesis, École Polytechnique Fédérale de Lausanne (EPFL), 2010. https://doi.org/10.5075/epfl-thesis-4714.

2. C.W.M. Robeller, Integral mechanical attachment for timber folded plate structures, thesis, École Polytechnique Fédérale de Lausanne (EPFL), 2015. https://doi.org/10.5075/epfl-thesis-6564.

3. C. Robeller, S.S. Nabaei, Y. Weinand, Design and fabrication of robot-manufactured joints for a curved-folded thin-shell structure made from CLT, in: W. McGee and M. Ponce de Leon (Eds.), Robot Fabrication in Architecture, Art and Design 2014. Springer International Publishing, New York, 2014: pp. 67–81. https://doi.org/10.1007/978-3-319-04663-1.

4. C. Robeller, Y. Weinand, Interlocking folded plate – Integral mechanical attachment for structural Wood panels, Int. J. Sp. Struct. 30 (2015) 111–122. https://doi.org/10.1260/0266-3511.30.2.111.

5. C. Robeller, Y. Weinand, Fabrication-aware design of timber folded plate shells with double through tenon joints, in: D. Reinhardt, R. Saunders, J. Burry (Eds.), Robot Fabrication in Architecture, Art and Design 2016. Springer International Publishing, New York, 2016: pp. 166–177. https://doi.org/10.1007/978-3-319-26378-6_12.

6. C. Robeller, A. Stitic, P. Mayencourt, Y. Weinand, Interlocking folded plate: Integrated mechanical attachment for structural Wood panels, in: P. Block, J. Knippers, N.J. Mitra, W. Wang (Eds.), Advances in Architectural Geometry 2014. Springer International Publishing, London, 2014: pp. 281–294. https://doi.org/10.1007/978-3-319-11418-7_18.

7. C. Robeller, P. Mayencourt, Y. Weinand, Snap-fit joints – CNC fabricated, integrated mechanical attachment for structural wood panels, in: D. Gerber, A. Huang, J. Sanchez (Eds.), ACADIA 2014 – Design Agency: Proceedings of the 34th Annual Conference of the Association for Computer Aided Design in Architecture. ACADIA, Los Angeles, 2014: pp. 189–198.

8. P. Vestartas, L. Pelletier, M. Nakad, A. Rezaei Rad, Y. Weinand, Segmented spiral using inter-connected timber elements, in: Procedings of the IASS Annual Symposium 2019, IASS, Barcelona, Spain, 2019: 1–8.

9. M. Deuss, A.H. Deleuran, S. Bouaziz, B. Deng, D. Piker, M. Pauly, ShapeOp—A robust and extensible geometric modelling paradigm, in: M. Thomsen, M. Tamke, C. Gengnagel, B. Faircloth, F. Scheurer (Eds.), Modelling Behaviour. Springer International Publishing, 2015: pp. 505–515. https://doi.org/10.1007/978-3-319-24208-8_42.

10. J. Gamerro, C. Robeller, Y. Weinand, Rotational mechanical behaviour of wood-wood connections with application to double-layered folded timber-plate structure, Constr. Build. Mater. 165 (2018) 434–442. https://doi.org/10.1016/J.CONBUILDMAT.2017.12.178.

11. C. Robeller, J. Gamerro, Y. Weinand, Théâtre vidy Lausanne — A double-layered timber folded plate structure, J. Int. Assoc. Shell Spat. Struct. 58 (2017) 295–314. https://doi.org/10.20898/j.iass.2017.194.864.

12. V. Baudriller, J. Gamerro, M. Jaccard, C. Robeller, Y. Weinand, eds., Le pavillon en bois du Théâtre de Vidy, 1st ed, EPFL Press, 2017.

13. Robert McNeel & Associates, Rhinoceros 3D website, 2018. Available at: https://www.rhino3d.com/.
14. D. Rutten, Robert McNeel and associates, Grasshopper 3D website, 2019. Available at: https://www.grasshopper3d.com/.
15. J. Gamerro, Development of novel standardized structural timber elements using wood-wood connections, École Polytechnique Fédérale de Lausanne (EPFL), 2020. https://doi.org/10.5075/epfl-thesis-8302.
16. J. Gamerro, J.F. Bocquet, Y. Weinand, A calculation method for interconnected timber elements using Wood-Wood connections, Buildings. 10 (2020)61. https://doi.org/10.3390/buildings10030061.
17. J. Gamerro, J.F. Bocquet, Y. Weinand, Experimental investigations on the load-carrying capacity of digitally produced wood-wood connections, Eng. Struct. 213: (2020) 110576. https://doi.org/10.1016/j.engstruct.2020.110576.
18. European Committee for Standardisation (CEN), CEN-EN 1995-1-1:2005+A1 - Eurocode 5: Design of timber structures – part 1-1: General – Common rules and rules for buildings, Brussels, 2008.
19. Annen SA website. Available at: http://www.annen.lu
20. C. Robeller, M. Konakovic, M. Dedijer, M. Pauly, A double-layered timber plate Shell — Computational methods for assembly, prefabrication and structural design, Adv. Archit. Geom. 2016: 104–122. https://doi.org/10.3218/3778-4.
21. C. Robeller, M. Konaković, M. Dedijer, M. Pauly, Y. Weinand, Double-layered timber plate shell, Int. J. Sp. Struct. 32 (2017) 160–175. https://doi.org/10.1177/0266351117742853.
22. A.C. Nguyen, P. Vestartas, Y. Weinand, Design framework for the structural analysis of free-form timber plate structures using wood-wood connections, Autom. Constr. 107 (2019) 102948. https://doi.org/10.1016/J.AUTCON.2019.102948.
23. European Committee for Standardization (CEN), EN 26891: Timber structures – Joints made with mechanical fasteners – General principles for the determination of strength and deformation characteristics, Brussels, Belgium, 1991.
24. A.C. Nguyen, Y. Weinand, Displacement study of a large-scale freeform timber plate structure using a total station and a terrestrial laser scanner, Sensors. 20 (2020) 413. https://doi.org/10.3390/s20020413.
25. A.C. Nguyen, A structural design methodology for freeform timber plate structures using wood-wood connections, École Polytechnique Fédérale de Lausanne (EPFL), 2020. https://doi.org/10.5075/epfl-thesis-7847.
26. A. Rezaei Rad, Mechanical characterization of Integrally-Attached Timber Plate Structures: Experimental studies and macro modeling technique, École Polytechnique Fédérale de Lausanne (EPFL), 2020. https://infoscience.epfl.ch/record/276874?ln=en

Chapter 4

Design optimization in timber plate structures

Yves Weinand

We have seen through Chapter 3 how the continuous investigation of global and local interconnecting geometries of timber fold structures can lead to various results and many technical and esthetic solutions. The optimization process of the eight presented case studies is performed through continuous adaption, essentially reacting to fabrication constraints. Those fabrication constraints are generally expressed then in geometrical terms or performed through geometrical modifications. The geometry of folds, for instance, has to be adapted as shown with repercussions also on global form considerations within the process. One could conclude that fabrication constraints became the most important parameter of the optimization process regarding the design of interconnected timber plate structures.

Material constraints also run into considerations. Available dimensions of timber plate panels have to be considered while processing the nesting outlet. In terms of waste management, it is crucial to optimize the interaction between early formal decisions on a global scale of the structure and its repercussions locally, as far as the geometries of connections came into play. The quality of the nesting process directly affects the quantity of unused material. For instance, the Chapel of Saint-Loup remains an outstanding and unique example as its waste material tends theoretically toward zero if we could use a panel the size of the chapel.

Montage constraints also played a significant role in developing large scale structures, such as the Vidy theater and the Annen SA project, since montage joints must be discussed and determined within a given logistic frame. We are not discussing the connections of the overall wooden structure to its basement, which in many cases stay as a logistically classical identifiable connection where the foundations are made out of concrete and the wooden structure is secured to it with steel plates and bolts. In the case of the Vidy theater, the contracting company tends clearly to define 33 independent prefabricated pieces. This decision led to two different wood-wood connections that were not initially part of the parametric model. One could argue that those connections added later in the design process are less pure and express another tectonic. Moreover, the assembly process itself differs also for those newly added connections. So, both criteria appear to alter the initial full coherence of the system and weakens it in part. In the case of the Annen SA project, montage joints within the structure had to be avoided altogether. Each arch is mounted horizontally as a whole, pivoted, and lifted toward its final position. One could say that a montage joint needed to be avoided, given the primary assembly sequence of the 'knitted' structure and that the geometrical topology of that montage joint would turn out to be difficult to execute since it would require more and varying connectors and harm the coherence of the primary assembly logistic. To conclude, fabrication constraints and montage constraints essentially dominate the design process, the execution logistics, and results. The underlying parametric model masters and includes all constraints and allows us to adapt and include essential geometrical choices.

So how could structural design decision-making processes be integrated into the workflow illustrated beforehand? All presented case studies of Chapter 3 also included their mechanical analyses. Mechanical models have been defined, and data flow has been described to facilitate finite element (FE) models and macromodels' conceptions as a newly defined way for mechanical characterizations of folded timber plate structures and their integrated connections. But those mechanical tools and models were implemented afterward. The main geomatical decisions were already made when the mechanical observation began. We did not break the current iterative realm, which divides the design process into separate categories or phases. Here the structural engineer 'jumps in' at a later stage.

Thus, one of the significant motivations regarding this book is to illustrate that structural design investigations within the context of the structural design of interconnected timber plate structures have not been adequately addressed or explored yet. The decision-making processes, as described here, have not considered mechanical observations at first. How would an early input of mechanical observations improve and optimize geometrical choices that need to be implemented into the parametric master model? How would those mechanical constraints help to define geometrical border conditions and thus 'compete' with fabrication and montage constraints, which, on their side, would require geometrical adaptations of another kind? It is fascinating to see how an architectural synthesis is needed in every construction project and how the architect needs to balance multiple contradictories and competing criteria to determine the best course of action. But what about the engineering approach and the mechanical or structural design synthesis of timber plate structures using wood-wood connections? We do not intend here to open the discussion on the interaction between form and structure. We do not intend to analyze how the architect and the structural engineer have to 'divide' their competence applied to both form and structure. But we would like to open a window onto the structural design process, which would move mechanical decision-making upstream and within the basic geometrical choices that underlie the design of the project. Timber plate structures allow for structural design choices that might be newly integrated in a holistic manner looking at the global picture and result. Within the frame of this book's title, we present two case studies respecting the described workflow by adding mechanical observations at an early stage.

4.1 PREFABRICATED TIMBER PLATES WITH PRECURVED GEOMETRY AND PRECAMBER EFFECT

Chapter 3 exposed in detail prefabricated timber plates with standard geometry. The double-layered orthotropic plate structure, its static height and span, the double-layered geometry and assembly of aisles and flanges made out of oriented strand board (OSB) panels, the number and the geometrical contours of tenons were all related to its primary purpose: propose a simple and recyclable shed structure of a new kind. In this case, it spans more than 10 meters. The final dimensioning criteria for such structures are always deflections. In this case, we need to limit deflections under permanent and death loads at the serviceability limit state (SLS). Gamerro [1] and Gamerro et al. [2, 3] dimensioning procedure allows for a fast and accurate dimensioning of this orthotropic plate structure, taking into consideration the semirigid behavior of its connections and panel disposal. Thus, we can determine the exact static deflection under such loads. Its value is 95 mm.

The precamber beam structure is a classical tool used to limit deflections. As the fabrication and assembly process defines major border conditions for such a structure, the precambered result could be achieved by reframing the geometry of the cutting process and the sequences of the montage of each part of the system. In this case, as a first step, we propose to drill a curved shape for all panels used as flanges in the system. The applied shape reproduces a curvature that corresponds to a ratio of 95mm over 10 meters. All panels are integrated into the system while flipped upward so the curvature is reversed and 'opposes' the upcoming deflections under permanent and death loads.

To introduce the aisle panels and fix them onto the curved shaped flange panels, we propose a second step to actively bend those aisle panels by imposing the same curvature. In fact, the aisle panels need to be held in that specific curved position and then be positioned on the top and the bottom of the flange panels. Still holding the impose curvature, the aisle panels in a third step then move using a simple translation into its final position. The structural system is shown in Figure 4.1a-c.

But how do those geometrical decisions affect the assembly process and the shape of each tenon? The reference project uses exclusively rectangular and orthotropic geometries for its main form as a rationalized construction system for standardized architecture. Wood-wood connections such as tenons and holes are drilled in respect of the same rectangular geometrical frame. Singular exceptions are mesh holes directly related to the cutting process. These mesh holes will also appear in this present proposal. When we observe Figure 3.12 of Chapter 3, we recognize under c.3 a conically shaped tenon linked to the top aisle. The conic shape comes from the need to facilitate the assembly process. The insertion vector is singular and vertical.

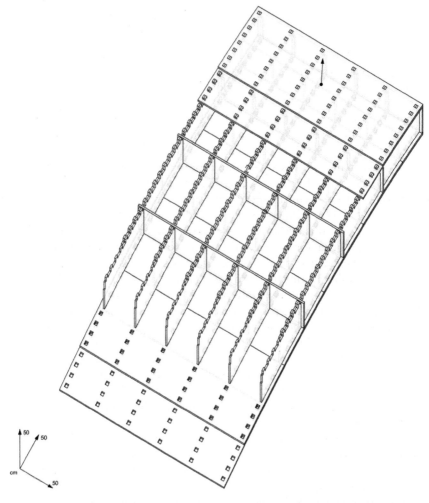

Figure 4.1 Prefabricated precurved timber plate with wood-wood connections. Project information: Robotic Hall Mobic, Harzé (Belgium). Building Type: Industrial hall. Timber Company: Mobic SA (Belgium). Robotic Company: IMAX PRO (Belgium). Architect: Yves Weinand, Liège (Belgium). Timber Engineering: Bureau d'Etudes Weinand, Liège (Belgium). Technology Transfer: Laboratory for Timber Construction, IBOIS, EPFL, Prof. Yves Weinand, Julien Gamerro, Martin Nakad, Nicolas Rogeau.

Research credit: Gamerro [1], and Gamerro et al. [2, 3].

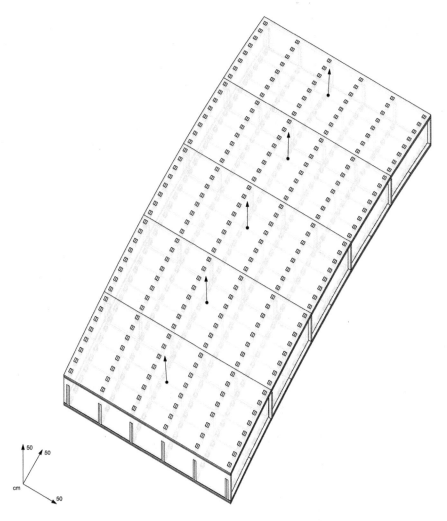

Figure 4.1 (Continued)

For the present case, we still have a vertical insertion vector to assemble the actively bent aisle panels with the preshaped curved flange panel. The tenon itself can still be of slight conical shape. Still, the vertical centerline of each tenon now needs to be parallel to the vertical insertion vector of the overall system. Thus, the centerline of each tenon will no longer be perpendicular to the curved centerline of the curved shaped flange panel. This has two main consequences: the shape of each tenon will be asymmetrical, and the shape of each tenon will be different.

The described development concerns the assembly process for the top aisle panel. By returning to Chapter 3's Figure 3.12, we see under d.2 that the conic shape has not been reproduced at the bottom to assemble the

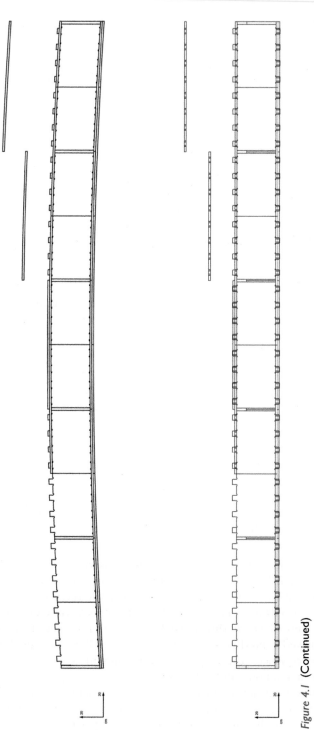

Figure 4.1 (Continued)

bottom aisle. Here, a distinct notch-shaped connection has been produced for each layer of the two aisle panels. Those notches are needed to hold the bottom aisle panels upward, resisting gravity. In contrast to the top part, where both aisle panels are introduced simultaneously by translation, the bottom aisle panels need to be introduced one by one.

In the present discussed case, both actively bent aisle panels must be held in their curved position and assembled one by one. The consequence of this demands a revision of each notch shape, which will undergo minor geometrical adaptions. Still, those changes need to be coded, nested, and finally cut. Once those geometrical observations are understood and integrated into the fabrication process, we end up producing a very efficient mechanical system. As mentioned before, the system will properly balance deflections under permanent and death load while finding a perfectly horizontal position. Creep could even be considered by increasing the value of the static deflections by a given creep factor and applying those additional deflections to the chosen precambered curvature. Notably, we achieve a system with an augmented mechanical capacity without additional production costs by simply manipulating its geometrical configurations. The coding and cutting process vary only slightly. In addition to achieving our initial goal, we also benefit from the following side effect: the very moment we relax the actively bent aisles panels, those panels tend to relocate into their initial horizontal position. By doing so, they will add a snap-fit effect onto each node or tenon, which will benefit the rigidity of each connection and the whole system.

4.2 DESIGN OPTIMIZATION IN DOUBLE-LAYERED DOUBLY CURVED FREEFORM INTEGRALLY-ATTACHED TIMBER PLATE STRUCTURES: OPTIMIZATION BY INTRODUCING CONTACT

The design optimization starts with probing into a new assembly method while satisfying the geometric tessellation of Integrally-Attached Timber Plate (IATP) structures. In other words, while timber plates keep their original topology and essentially remain quadrilateral, a new tiling method is introduced to direct forces in a way that they equally distribute stress on the wood-wood connections. As a result, the overall appearance and global architectural design of the model are kept, while the structural performance is enhanced. The structural behavior improvement is reflected in the relatively uniform distribution of loads to the wood-wood connections and the fact that the failure path recognized in the first benchmark is blocked. In this chapter, a new assembly corresponding to the double-layer IATP structure – hereafter, the optimized IATP structure – is introduced, and its structural performance is compared with the initial benchmark designed in Chapter 3. It is worth noting that the design tools developed and formulated in Chapter 2 are essentially valid in the current chapter. The performance of the optimized and initial prototypes is compared in terms of their structural displacement and the load-deformation curve of the midspan for each structure.

4.2.1 Design alternatives

Similar architectural design and geometry generation methodology used in the initial benchmark is adopted for the optimized IATP structure. As such, a similar surface segmentation process was used, and the optimized structure consists of top, bottom, cross-longitudinal, and cross-transverse plates. Moreover, the adaptation of the optimized tiling method ensures that the target surface assumed at the very beginning of the design process remains intact. However, the difference between the optimized and initially designed benchmark is that the parallelogram herringbone pattern used for the initial benchmark is shifted. This shift puts the top plates of a strip in between the connection line of the top plates of the neighboring strip. This shifting pattern was mainly designed due to the fact that the escape path (failure pattern) was appearing at the stretched bottom layer of the structure even if geometrically joinery is blocking the assembly sequence. In fact, the failure path appeared during loading testing and contradicted the initial assumption of the block graph. Geometrically, the structure is blocked due to connections, but the joints themselves break along one vector per escape path. Consequently, further studies were carried out to understand the application of the alternative structural system.

Previous studies have mostly focused on connections with sufficient load-bearing capacities to be used not only in timber frame structures but also with timber plates. The project framework requires keeping the same global shape while changing the nonuniform rational basis spline (NURBS) discretization pattern. The influence of the geometry patterns used in the assembly system on the global structural performance is investigated for the same case study. The goal is to modify the global geometry to enhance its mechanical behavior without changing the shape of shells.

The reference structural system, which uses the initial herringbone pattern, is illustrated in Figure 4.2a. Each shell is composed of an assembly of hexahedra-shaped boxes in this system, each made of two vertical and two horizontal plates. The latter forms the top and bottom layers of the structure. This structural system has been widely investigated in Chapter 2 and Chapter 3. Neighboring horizontal plates are connected with one degree of freedom (1DOF) through-tenon wood-wood connections through the cross plates, which are themselves assembled with dovetail joints. Each box shares its vertical panels with neighboring boxes. Boxes are individually formed. Afterward, each box component is inserted along the insertion vector, defined by the direction of the remaining DOF of the connections. As an alternative system, a hexagonal pattern is proposed where continuous

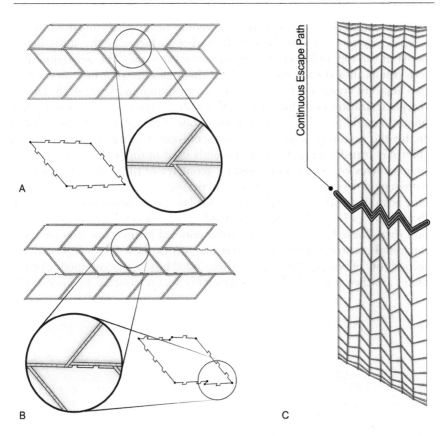

Continuous Escape Path

Figure 4.2 Changes in the design pattern of double-layered double-curved timber freeform plate structure from (a) to (b) to avoid continuous failure pattern (escape path).

Research credit: A.C. Nguyen, 2020 [4].

failure paths are interrupted, including the assembling sequence of individual components. The shape of the hexagon graphically resembles the same quad elements and gains an advantage for considering structural performance versus the built project. The geometry is shown in Figure 4.2b.

4.2.2 Shifted herringbone pattern and contact zones

Double-layered double-curved timber plate shells have been achieved using the same computer-aided design (CAD) programming using Rhinoceros, the Grasshopper interface, and C# programming language. A hexagonal tessellation was applied to the same double-curved surface. In this study, both structural systems were investigated regarding mesh discretization, timber plate generation, assembly sequence, and engineering calculation. The shifted hexagonal pattern was investigated to challenge the problems encountered in the initial system. The geometrical constraints used for the reference system are kept, where the target surface and the static height between the top and bottom plates remain the same. The hexagonal pattern results in a shifted pattern where the long edge of a box meets with two adjacent components instead of one, as shown in Figure 4.2b. Due to the planarization and thickness of the plates, each top and bottom plate obtains an additional edge. The shape of horizontal plates was modified to enable the assembly sequence without collision. Overall, the main design change is transforming the quadrilaterals used in the herringbone pattern to nonconvex octagons. The resulting geometry of the shifted herringbone pattern modifies the assembly sequence of the system, as is shown in Figure 4.3. One possible solution is to assemble quadrilateral and nonconvex octagonal-shaped plates by adding additional side plates depending on an even or odd row. These four-sided boxes are assembled in the same manner as in the initial construction system. After one row of these boxes is placed, the second row of boxes, consisting of three-sided boxes, is positioned. Vertical plates are then inserted along the assembly vector directed by joint angle.

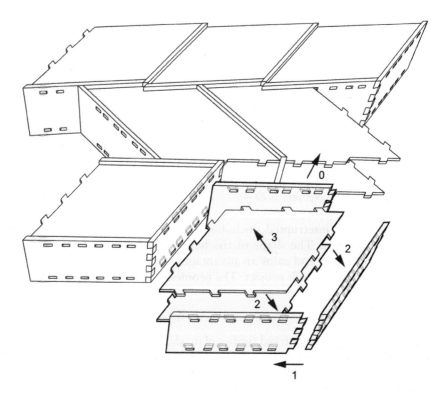

Figure 4.3 The assembly sequence of components follows a similar box component topology introducing additional side plates depending on an even or odd row.

Research credit: A.C. Nguyen, 2020 [4].

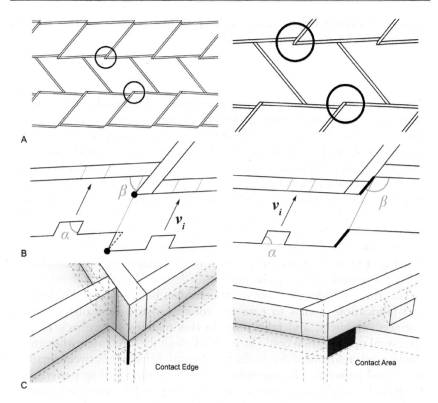

Figure 4.4 Angles of the top and bottom plates (a) angles in question, (b) left figure requires geometrical modification, right no collision, (c) the change in convex corners results in contact edge while the concave edge has the advantage of the full contact area.

Research credit: A.C. Nguyen, 2020 [4].

The same process is repeated for subsequent rows. Consequently, the proposed assembly sequence requires more steps and is, therefore, more complicated than the initial assembly sequence.

The introduction of nonconvex octagonal-shaped plates with the shifted herringbone pattern has the advantage of providing zones of abutment of the horizontal plates in the structural system, as is shown in Figure 4.4a-c. As discussed before, the shape of the plates has to be modified for abutment angles beta (β) less than tenon angles alpha (α), defined by the insertion vector v_i and unique for each edge of the plates, to ensure their insertion. The kink in the hexagonal plate cannot block the insertion; therefore, a small gap is introduced. For abutment angles β larger than tenon angles α, no modification of the plate shape was required to see the difference between angles in Figure 4.4b. Therefore, the contact zone remained the full area of the abutment, as shown in Figure 4.4c. According to the assembly sequence illustrated in Figure 4.3, the insertion vector of the vertical plates is identical to the insertion vectors of the horizontal plates with which three-sided boxes are formed.

4.2.3 Polygonal mesh datastructure – NGon

The new geometrical development conceptually follows the initial design, while the algorithm is built differently following polygonal mesh data-structure using NGon plugin for Rhino and Grasshopper[1]. The quad or triangle mesh datastructure has in-build adjacency queries such as edge-to-face, face-to-face, faces-to-vertex; however, the hexagonal pattern requires a software modification in terms of using polygonal mesh datastructure. Internally, the polygonal meshes are stored as triangles groups and graphically visualized as one entity displaying only the outer boundary of the mesh face. Previously described fabrication data, such as outlines of plates, requires identification of adjacent mesh faces to polygon faces. The plane offset and plane-to-plane intersection methodology are kept similar and are implemented following the previously described joinery solver.

This time, the design framework considers both structural calculation and fabrication models to ease the collaborative data transfer between engineer, architect, and fabricator. In this manner, a feedback loop for structural optimization could be enabled. Both the CAD algorithm and FE model needs refinement due to the topological difference between quad and hexagonal patterns. The algorithm starts by applying tiling patterns to a NURBS. Using the target surface shown in Figure 4.5a, pairs of hexagonal nodes are positioned at U and V NURBS subdivision and form a shifted herringbone pattern (see Figure 4.5b). Since the mesh is doubly curved, planarization has to be applied to the overall geometry. The advantage of the structure is that it contains a large number of self-similar elements that are relatively small in relation to the whole shape of the arch. In this study, angles were sufficiently low such that enough joinery areas were left for the wood-wood connections. As a result, polygons can be projected to an average plane of each hex-polylines (see Figure 4.5c). The shape of the projected polylines is slightly shifted within neighbor elements. As a consequence, this satisfies both fabrications and structural constraints without a need for secondary planarization procedures. Furthermore, edges of extruded mesh faces must be planar and equal along border lines to fit foundation and glass details. The plane projection and intersection method thus remain the same following polygonal mesh adjacency.

Insertion vectors have to be defined for each box, such that plate edges forming an obtuse angle share a unique insertion vector (see Figure 4.6a) for the bottom layer of the shell structure. Tenon polylines are inserted into closed plate polylines, whereas the female outlines are added to side outlines. The data structure, called Outline, is described as a list of polygons

---

[1] https://www.food4rhino.com/app/ngon

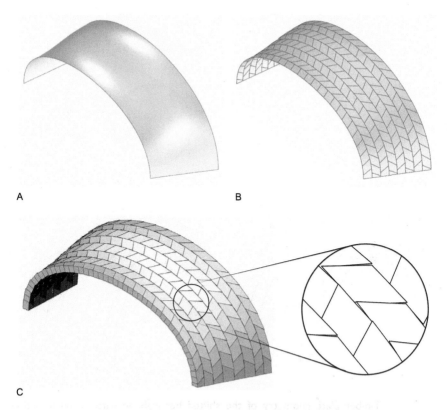

A   B

C

Figure 4.5 Hexagonal tiling to obtain planar hexagonal boxes: (a) target surface, (b) hexagonal subdivision, (c) planarization based on plane-to-plane intersection per each NGon mesh edge with top and bottom planes obtained by fitting hexagonal points.

Research credit: A.C. Nguyen, 2020 [4].

with joinery segments. Each timber plate is described using a pair of outlines that contain male and female polygons depending on the joinery. The side joints remain the same using dovetail connections. Next, the algorithm generates male and female joint segments and assembles closed polylines to form one continuous polygon. For visualization purposes, polylines are meshed using an ear-clipping algorithm that can consider holes as well.

Given that the mesh topology follows a shifted herringbone pattern, the performance comparison between the reference and optimized structures became possible. The geometry required for structural calculation is generated by the algorithm developed in order to perform the FE analysis. The methodology was applied to the arches shown in Figure 4.6 since the NURBS geometry is self-similar. The reference and alternative systems are compared for assemblies of 25 and 28 boxes, respectively, using the semi-rigid spring model developed in Chapter 2.

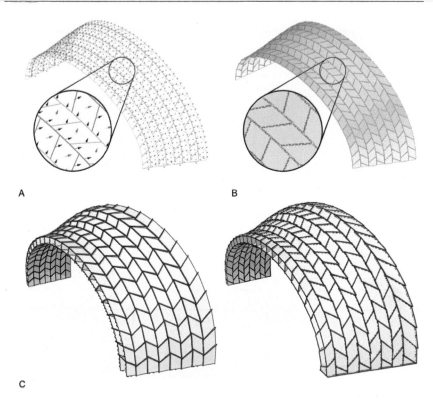

Figure 4.6 Timber plate geometry of the shifted herringbone pattern: (a) insertion
vectors, (b) tenons for through tenon-joint derived from insertion vectors,
(c) side-by-side comparison between quad and hexagon subdivision.

Research credit: A.C. Nguyen, 2020 [4].

4.2.4 Engineering analysis

Using the modeling procedure described in Chapter 2, the CAE FE model is
generated. The model, primarily investigated by Nguyen [4] and Nguyen
et al. [5], was built in ABAQUS™, version 6.12, using the scripting interface. It is worth noting that the custom scripting code automating the generation of the FE model was modified to consider the plates' non-octagonal
shape. The contact between the timber plates used springs in a series. The
spring elements were infinitely rigid in compression and free under tensile
behavior. The geometry of the reference and alternative models used in the
FE analysis is shown in Figure 4.7a-b, respectively.

A distributed load of 2 kN/m^2 was applied to the top layer of the systems. The results indicate that the alternative system helped avoid the
continuous failure paths, which appeared at the bottom stretched layer
of the initial system by organizing boxes in staggered rows. New methods were provided to integrate the generation of the suggested pattern

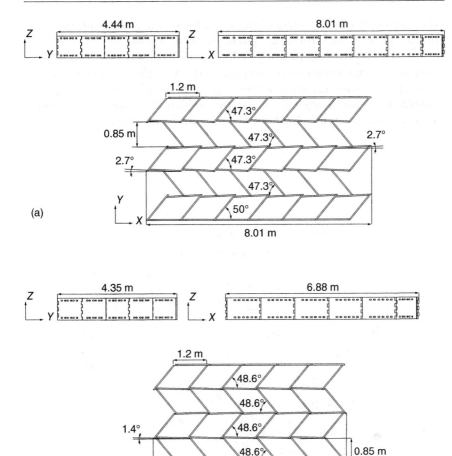

Figure 4.7 (a) Geometry of the 5×5 specimen considered for numerical investigations applying the initial herringbone pattern (the reference model). The same geometric parameters as in the tested prototype of 5 × 3 boxes, studied in Chapter 3, were considered, (b) geometry of the 28-box specimen was considered for numerical investigations applying the shifted herringbone pattern (the optimized alternative).

Research credit: Nguyen et al [5]. and A.C. Nguyen, 2020 [4]

in the design and structural calculation frameworks by close collaboration between architect and engineer. Based on numerical investigations, the assembly pattern was shown to significantly influence the studied structure's performance. The results of both the optimized and original structures are shown in Figure 4.8b-d and Figure 4.8e-f, respectively. The proposed design, involving additional abutment areas of the boxes, was shown to enhance the interlocking assembly of the plates and, therefore, the stiffness of the structural system by 76%. Tensile forces in the connections were also significantly reduced, increasing the distributed load to reach the maximum load-carrying capacity of the connections in tension to 85%. Instead, shear forces appear at the plate edges around the abutments when using the alternative structural system. The contact areas shown in Figure 4.4 have the potential to increase the structural

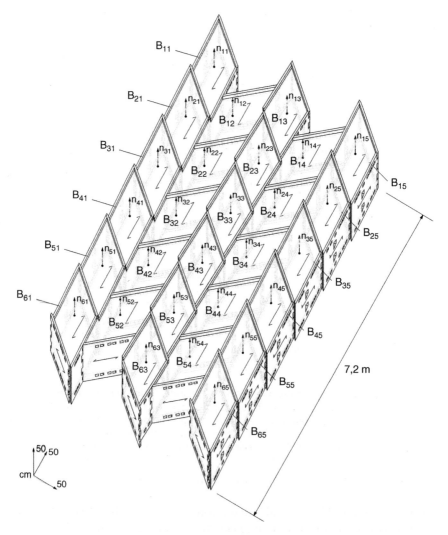

Figure 4.8 (a) Geometry of the optimized structure and associated (b) vertical displacement and (c-d) von Mises stress, and (e) vertical displacement and (f) von Mises stress associated with the reference structure.

Research credit: A.C. Nguyen, 2020 [4].

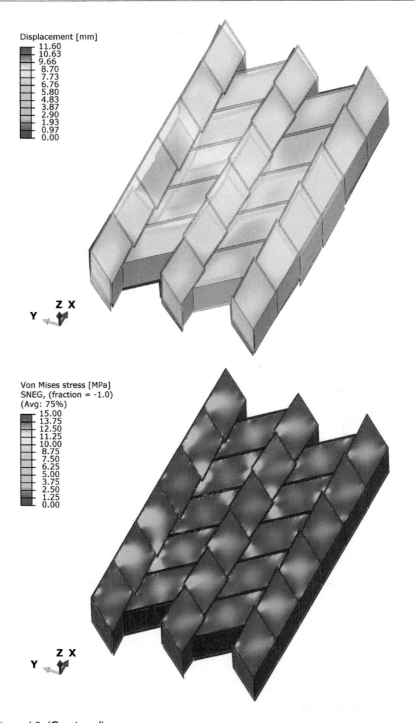

Figure 4.8 (Continued)

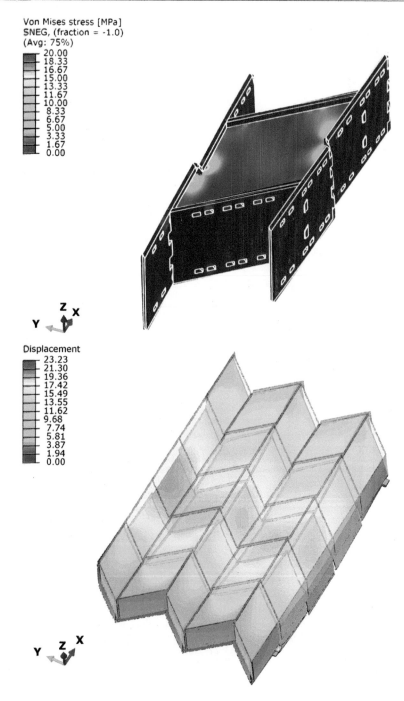

Von Mises stress [MPa]
SNEG, (fraction = -1.0)
(Avg: 75%)
20.00
18.33
16.67
15.00
13.33
11.67
10.00
8.33
6.67
5.00
3.33
1.67
0.00

Displacement
23.23
21.30
19.36
17.42
15.49
13.55
11.62
9.68
7.74
5.81
3.87
1.94
0.00

Figure 4.8 (Continued)

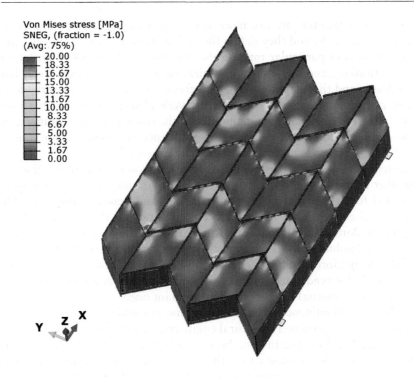

Von Mises stress [MPa]
SNEG, (fraction = -1.0)
(Avg: 75%)

Figure 4.8 (Continued)

performance, too. Besides the structural optimization, the study shows a need to integrate structural design in the first stages of geometrical development by rethinking existing data structures that could serve both engineers and architects. Given that the abutment areas were shown to have the potential to further enhance the structural performance of the alternative system, it is concluded that the shifted herringbone pattern would have to be applied on full arches to accurately assess their structural improvement.

CONCLUSION

This book exposes a new kind of structure. The traditional separation between primary and secondary structural bearing elements disappears throughout all presented case studies. But we also exposed a new type of structural thinking applied to those structures. Both presented case studies tend to illustrate a revival of structural design thinking. The system evolves as a whole since its genetic base is a parametric model in which parameters of different kinds can be easily defined. The two presented case studies show how parameters of a geometric nature can lead to mechanical performances.

But those parameters are not more separated from the architectural body, they are part of it, and they define the architectural body in its essence and style. This is of particular interest also for structural engineers. Structural design thinking and mechanical optimization processes may flow freely into the design and the adaption of the constructed model. As shown, the force flow can be controlled and modified. Overloaded areas of the structural envelope can be unloaded by manipulating the envelopes' inherent inertia in those zones. By doing so, the force flow will dislocate and search for another path guiding those loads toward more rigid adjacent zones.

Integrated mechanical attachments or wood-wood connections allow for another aspect — the potential controlled separation of each of the three internal forces: normal forces, shear forces, and bending moments. The individual geometric design of each tenon allows for the selection of one of the predominant internal forces to be absorbed as a primary force. So the overall solution will be a better distribution or a distinct distribution of all internal forces to be taken. Wood-wood connections can be designed specifically for predefined internal forces to be taken, and each connection can take one internal force as the predominant one.

The role and influence of structural engineers could revive. Of course, we have seen how renowned structural engineering companies also take up the lead regarding building information models in construction management constellations. But in most cases, the essence of the constitutive quality of a given building information model is not questioned. Its properties are not pushed further, and the quality of those BIM models stays at the state-of-the-art level without further development.

The separations between general static construction documents and detailing disappear. Secondary and primary bearing systems and construction detailing execution documents are replaced by one holistic system that defines space, structure, and envelopes. The system evolves as a whole since its genetic base is a parametric model from which all construction partners may benefit.

Showing a constitutive evolution of what a BIM model could become as an evolving parametric tool set, the structural engineer's input might add to the decision process upstream during the design process. Over the last century, the importance and the input of structural engineers have been reduced. Part of the engineer's input was reduced to the solving of small-scale problems. Introducing and enhancing a more active role for structural engineers is a more fertile discussion for contemporary architectural construction systems in sustainable considerations.

REFERENCES

1. J. Gamerro, Development of novel standardized structural timber elements using wood-wood connections, École polytechnique fédérale de Lausanne (EPFL), 2020. https://doi.org/10.5075/epfl-thesis-8302.

2. J. Gamerro, J.F. Bocquet, Y. Weinand, A calculation method for interconnected timber elements using wood-wood connections, Buildings. 10 (2020) 61. https://doi.org/10.3390/buildings10030061.

3. J. Gamerro, J.F. Bocquet, Y. Weinand, Experimental investigations on the load-carrying capacity of digitally produced wood-wood connections, Eng. Struct. 213 (2020) 110576. https://doi.org/10.1016/j.engstruct.2020.110576.

4. A.C. Nguyen, A structural design methodology for freeform timber plate structures using wood-wood connections, École polytechnique fédérale de Lausanne (EPFL), 2020. https://doi.org/10.5075/epfl-thesis-7847.

5. A.C. Nguyen, B. Himmer, P. Vestartas, Y. Weinand, Performance assessment of double-layered timber plate shells using alternative structural systems, Proceedings of the IASS Symposium 2019 – Structural Membranes, 2019, pp. 2919–2926. http://infoscience.epfl.ch/record/273554.

Index